Lecture Notes in Mathematics

1550

A.A. Gonchar   E.B. Saff   (Eds.)

Methods
of Approximation Theory
in Complex Analysis
and Mathematical Physics

Euler Institute, 1991

Springer-Verlag

Lecture Notes in Mathematics 1550

Editors: A. Dold, Heidelberg
          B. Eckmann, Zürich
          F. Takens, Groningen
Subseries: POMI and Euler International Mathematical
           Institute, St. Petersburg
Adviser: L. D. Faddeev

Editors

Andrei A. Gonchar
The Steklov Mathematical Institute
Russian Academy of Sciences
Vavilova 42
Moscow, GSP-1, 117966, Russia

Edward B. Saff
Institute for Constructive Mathematics
University of South Florida
Tampa, FL 33620, USA

Publisher's Note:
This book was originally printed in 1992 for the Euler Institute,
St. Petersburg, by Nauka Publishers, Moscow, then reissued by
Springer-Verlag in 1993, under a new cover, in the Lectures Notes
in Mathematics series. This circumstance explains the slight differences in appearance (format etc.) from the rest of the LNM
series.

Mathematics Subject Classification (1991): 30E10, 41A21,
41A46, 41A50, 41A65, 42C05, 42C15, 33A65, 33A70, 31C15

ISBN 3-540-56931-6
Springer Verlag Berlin Heidelberg New York
ISBN 0-387-56931-6
Springer-Verlag New York Berlin Heidelberg

46/3140-543210 - Printed on acid-free paper

Russian Academy of Sciences
The Euler International
Mathematical Institute

METHODS
OF APPROXIMATION THEORY
IN COMPLEX ANALYSIS
AND MATHEMATICAL PHYSICS

Leningrad, May 13-24, 1991

Editors
A.A.Gonchar, E.B.Saff

MOSCOW "NAUKA" 1992

This book is a collection of selected papers written by participants of international seminars on "Methods of Approximation Theory in Complex Analysis and Mathematical Physics". The seminars were organized in Leningrad, May 13–26, 1991 by the Euler International Mathematical Institute of Russian Academy of Sciences (St. Petersburg, Russia) and Institute for Constructive Mathematics of University of South Florida (Tampa, Florida).

This meeting was the second part of a joint scientific programme of Academy of Sciences and of National Science Foundation. The first conference "Approximation Theory" took place at Tampa, Florida, March 19–23, 1990. The results of this conference were published by Springer-Verlag (A.A. Gonchar, E.B. Saff, Progress in Approximation Theory, An International Perspective, Springer series in Computational Mathematics, 19, 1992).

The present book supplements the results lectured at Florida by new developments occured in the fields of common interests of international scientific groups participated in this programme. The editors included papers on orthogonal polynomials, wavelets analysis, constructive approximation theory, function theory in the unit disc, theory of n-widths, and other topics.

The seminars and publication of this book were financially supported by the Euler International Mathematical Institute (EIMI), Russian Academy of Sciences, NSF, and UNESCO.

The camera-ready of the book was prepared by the Electronic Publishing Center of the EIMI.

Methods of Approximation Theory
in
Complex Analysis and Mathematical Physics

May 13–26, 1991   Leningrad

International Organizing Committee
A.Gonchar, E.Saff, R.DeVore, S.Khrushchev

List of Participants

Aptekarev Alexander	Moscow
Askey Richard A.	Madison
Atakishiyev Natic	Baku
Bagby Thomas	Bloomington
Belyi Vladimir I.	Donetsk
Birman Michail	St.Petersburg
Borichev Alexander	St.Petersburg
Brennan James	Lexington
Brudnyi Yuri	Yaroslavl
Chui Charles	College Station Texas
Erdelyi Tamas	Columbus Ohio
Faddeev Ludwig D.	St.Petersburg
Geronimo Jeffrey	Atlanta
Gonchar Andrei	Moscow
Gulisashvili Archil	Tbilisi
Havin Victor	St.Petersburg
Ismail Mourad E-H.	Tampa
Ivanov Oleg	Donetsk
Kaliagyn Valery	Nizhnii Novgorod
Kashin Boris	Moscow
Khruchshev Sergei	St.Petersburg
Kolyada Victor	Odessa
Levin Alexei(Eli)	Tel-Aviv
Li Xin	Orlando
Maimeskul Victor	Donetsk
Micchelli Charles	Yorktown Heights

Nevai Paul	Columbus
Newman Donald J.	Temple
Oskolkov Konstantin	Moscow
Parfenov Oleg	St.Petersburg
Pekarskii Alexandr A.	Grodno
Pritsker Igor E.	Donetsk
Rakhmanov Eugene	Moscow
Rivlin Theodore	Yorktown Heights
Rybkin Alexei	St.Petersburg
Saff Edward B.	Tampa
Shekhtman Boris	Tampa
Shirokov Nikolai	St.Petersburg
Shulimanov Peter	Sverdlovsk
Solomyak Michail Z.	St.Petersburg
Stahl Herbert	Berlin
Suetin Sergei	Moscow
Suslov Sergei	Moscow
Temlyakov Vladimir	Moscow
Tikhomirov Vladimir	Moscow
de Vore Ronald	Columbia

CONTENTS

CONTENTS

BERNSTEIN THEOREMS
FOR HARMONIC FUNCTIONS

Thomas Bagby and Norman Levenberg

1. Introduction

A famous theorem of Bernstein states that a bounded function f on the interval $[-1, 1]$ can be extended to a holomorphic function on a complex open neighborhood of $[-1, 1]$ if and only if the sup-norm distance from f to the holomorphic polynomials of degree $\leq n$ decays exponentially with n. The following definitive result of this type follows from the results in Walsh [16, Chapter IV].

THEOREM 1.1 (Bernstein-Walsh). *Let K be a compact subset of $\mathbf{C}$ such that $\mathbf{C} \backslash K$ is connected and regular for the Dirichlet problem. Let f be bounded on K, and for each nonnegative integer n set*

$$D_n(f, K) = \inf\{\|f - p\|_K : p \text{ is a holomorphic polynomial of degree } \leq n\}.$$

Let $0 \leq \rho < 1$ Then we have $\limsup_{n \to \infty} D_n(f, K)^{1/n} \leq \rho$ if and only if f is the restriction to K of a function holomorphic on $\{z \in \mathbf{C} : g_K(z) < \log 1/\rho\}$.

Here and in the future we use the notation $\|\varphi\|_K \equiv \sup_K |\varphi|$; the Green function g_K is the continuous function on $\mathbf{C}$ which is harmonic on $\mathbf{C} \backslash K$, vanishes on K, and is equal to $\log |z|$ plus a bounded function near infinity. Note that this theorem includes a quantitative version of Runge's theorem for polynomial approximation. Extensions of Theorem 1.1 to $\mathbf{C}^N$ have been given in [13], [4], [19], [15], [5].

In this paper we consider analogous results for harmonic functions in $\mathbf{R}^N$. For each nonnegative integer n we define the vector space $\mathcal{H}_n$ of all real polynomials of degree $\leq n$ which are harmonic functions on $\mathbf{R}^N$; for each compact set $K \subset \mathbf{R}^N$, and each bounded function f on K, we let

$$d_n(f, K) = \inf\{\|f - p\|_K : p \in \mathcal{H}_n\}.$$

Our goals are to prove that for a harmonic function f on an open neighborhood of K the distances $d_n(f, K)$ decay exponentially with n and to prove results of converse type. The most general results we obtain are given in Theorems 3.1 and 4.1. In particular we prove the following harmonic Bernstein theorem with simple geometric hypotheses.

THEOREM 1.2 (Bernstein theorem for harmonic functions). *Let K be a compact subset of $\mathbf{R}^N$ such that every point of ∂K is analytically accessible from $\operatorname{int} K$. Suppose that $\mathbf{R}^N \backslash K$ is connected. Let f be a bounded real-valued function on K. Then f extends to a harmonic function on an open neighborhood of K if and only if*

$$\limsup_{n \to \infty} d_n(f, K)^{1/n} < 1. \tag{1}$$

Here ∂K and $\operatorname{int} K$ refer to the boundary and interior of K in $\mathbf{R}^N$. The statement that a point $a \in \partial K$ is *analytically accessible* from $\operatorname{int} K$ means that there exists a real-analytic function $h : (-1, 1) \to \mathbf{R}^N$ such that

$$h((0,1)) \subset \operatorname{int} K \quad \text{and} \quad h(0) = a. \tag{2}$$

If K is the closure of a bounded domain in $\mathbf{R}^N$ with C^1 boundary, then every point of ∂K is analytically accessible from $\operatorname{int} K$ since we may take h to be linear.

In case the dimension $N = 2$, the results of the present paper follow from work of Walsh [18], Siciak [14], and Nguyen Thanh Van and B. Djebbar [11] (see Section 4 below); however, the proofs of these results in the case $N = 2$ are all dependent on the analogous theory for holomorphic functions in $\mathbf{C}$, and hence give no indication of the situation in higher dimensions. Our proof of Theorem 1.2 will depend on the development of certain Schwarz lemmas for holomorphic functions of several complex variables, which are applied to harmonic functions suitably extended to complex domains; these ideas are developed in Section 2, and then applied in Section 3 to show that a function which is harmonic on an open neighborhood of a compact set K must satisfy (1). Methods from the theory of functions of several complex variables are also used in Section 4 to show that the converse is true under certain hypotheses on K. Our presentation will be self-contained: except for elementary properties of holomorphic and plurisubharmonic functions, we prove the facts concerning functions of several complex variables which we need. We mention that our techniques may also be used to prove similar results for solutions to certain elliptic equations.

In this paper we will say that a real-valued function on a compact set $K \subset \mathbf{R}^N$ is harmonic provided it can be extended to a harmonic function on an open neighborhood of K in $\mathbf{R}^N$. If $r > 0$ we use the notation $B(a, r) = \{x \in \mathbf{R}^N : |x - a| < r\}$ for balls in $\mathbf{R}^N$ and $\tilde{B}(a, r) = \{z \in \mathbf{C}^N : |z - a| < r\}$ for balls in $\mathbf{C}^N$. Multi-index notation will be used throughout the paper; if $\alpha = (\alpha_1, \ldots, \alpha_N)$, then D^α indicates $\partial^{|\alpha|}/\partial x_1^{\alpha_1} \cdots \partial x_N^{\alpha_N}$ or $\partial^{|\alpha|}/\partial z_1^{\alpha_1} \cdots \partial z_N^{\alpha_N}$. We let λ denote Lebesgue measure.

This research was conducted while the second author was visiting Indiana University in the summer of 1991; both authors would like to thank the mathematics department. Special thanks are due to Eric Bedford; without his valuable conversations this paper would not have been possible.

2. Schwarz lemmas

An essential ingredient in our work is a type of Schwarz lemma for harmonic functions on domains in $\mathbf{R}^N$. Schwarz lemmas for harmonic functions on balls can be proved by spherical harmonic expansions [2, Lemma 2.4]. In the present generality our proofs will depend on first establishing a Schwarz lemma for holomorphic functions of several complex variables; some other Schwarz lemmas in this setting are given in [7].

If Ω is a bounded domain in $\mathbf{C}^N$ and $a \in \Omega$, we define the *pluricomplex Green function* for Ω with logarithmic pole at a as the function

$$g_\Omega(z,a) = \sup_v v(z),$$

where the supremum is taken over the family of all nonpositive plurisubharmonic functions v on Ω such that $v(z) - \log|z - c|$ has an upper bound near a. (The boundedness of Ω assures that this family contains the function $\log(|z - a|/R)$ for some $R > 0$, and hence is nonempty.) The pluricomplex Green function was introduced by Klimek [9] and is discussed in detail in [6] and [10], but we need only some elementary facts which we will prove here. From the definition we see that $g_\Omega(\cdot,a) \leq 0$ on Ω, and we obtain the *monotonicity property:* if $a \in \Omega \subset \Omega'$, then $g_{\Omega'}(\cdot,a) \leq g_\Omega(\cdot,a)$ on Ω.

LEMMA 2.1. *For the complex open ball $\tilde{B} = \tilde{B}(a,r)$ we have*

$$g_{\tilde{B}}(z,a) = \log \frac{|z - a|}{r}, \qquad z \in \tilde{B}.$$

Proof. It suffices to prove the result for $a = 0$, $r = 1$. From the definition we obtain $g_{\tilde{B}}(z,0) \geq \log|z|$. For the reverse inequality, let $v \leq 0$ be plurisubharmonic in Ω such that $v(z) - \log|z|$ has an upper bound near 0. Applying the removable singularities theorem for subharmonic functions, we may regard $v(z) - \log|z|$ as a subharmonic function in $\tilde{B}$, and hence from the maximum principle we see that $v(z) - \log|z| \leq 0$ for $z \in \tilde{B}$; thus $g_{\tilde{B}}(z,0) \leq \log|z|$, which completes the proof.

THEOREM 2.2. *Let Ω be a bounded domain in $\mathbf{C}^N$ and $a \in \Omega$.*

(i) The pluricomplex Green function $g_\Omega(\cdot,a)$ is a nonpositive plurisubharmonic function on Ω such that $g_\Omega(\cdot,a) - \log|z - a|$ is bounded near a.

(ii) If f is any holomorphic function on Ω satisfying $|f| \leq 1$ there, and $D^\alpha f(a) = 0$ whenever $|\alpha| < n$, then $|f| \leq \exp[n\, g_\Omega(\cdot,a)]$ in Ω.

Proof. (i) Let positive numbers $r < R$ be selected so that $\tilde{B}(a,r) \subset \Omega \subset \tilde{B}(a,R)$. From the monotonicity property and the preceding lemma we see that for each $z \in \tilde{B}(a,r)$ we have $\log|z - a|/R \leq g_\Omega(z,a) \leq \log|z - a|/r$.

From the theory of plurisubharmonic functions it follows that the upper semicontinuous regularization $g_\Omega^*(z, a) \equiv \limsup_{\zeta \to z} g_\Omega(\zeta, a)$ is plurisubharmonic in Ω. We have $g_\Omega^*(\cdot, a) \leq 0$ on Ω, and from the preceding paragraph we see that $g_\Omega^*(z, a) \leq \log |z - a|/r$ for $z \in \tilde{B}(a, r)$. We conclude that the function $v \equiv g_\Omega^*(\cdot, a)$ is one of the competing functions in the definition of $g_\Omega(\cdot, a)$, and hence $g_\Omega^*(\cdot, a) \leq g_\Omega(\cdot, a)$ in Ω; since the reverse inequality is obvious, we see that $g_\Omega(\cdot, a) \equiv g_\Omega^*(\cdot, a)$ is plurisubharmonic on Ω.

(ii) The conditions on f imply that there is a constant $C > 0$ such that $|f(z)| \leq C|z - a|^n$ for all z near a. Thus the function $v \equiv \frac{1}{n} \log |f|$ is admissible in the definition of $g_\Omega(\cdot, a)$, and (ii) follows.

COROLLARY 2.3 (Schwarz lemma for holomorphic functions). *Let Ω be a bounded domain in $\mathbf{C}^N$ and $a \in \Omega$. Let K be a compact subset of Ω. If f is any holomorphic function on Ω satisfying $|f| \leq 1$ there, and $D^\alpha f(a) = 0$ whenever $|\alpha| < n$, then $\max_K |f| \leq \rho^n$, where*

$$\rho \equiv \sup_K \exp[g_\Omega(\cdot, a)] < 1.$$

The fact that $\rho < 1$ follows from the fact that $g_\Omega(\cdot, a)$ is subharmonic and nonpositive in Ω. The rest of the corollary follows from Theorem 2.2.

We next use Corollary 2.3 to prove a Schwarz lemma for solutions of elliptic partial differential equations. Let $P(D) = \sum_{|\alpha| \leq m} a_\alpha D^\alpha$ be a partial differential operator with constant coefficients in $\mathbf{R}^N$ which satisfies the ellipticity condition that $\sum_{|\alpha| = m} a_\alpha \xi^\alpha = 0$ only for $\xi = 0$. We recall the Weyl lemma which states that a distribution f satisfying $P(D)f = 0$ on an open set must be a C^∞ function satisfying $P(D)f = 0$ in the classical sense.

THEOREM 2.4 (Schwarz lemma for solutions of elliptic equations). *Let Ω be a domain in $\mathbf{R}^N$, and $a \in \Omega$. Let K be a compact subset of Ω. Then there exist constants $C > 1$ and $\rho \in (0, 1)$, depending only on Ω and K, with the following property. If f is a solution of $P(D)f = 0$ on Ω satisfying $|f| \leq 1$ there, and $D^\alpha f(a) = 0$ whenever $|\alpha| < n$, then $\max_K |f| \leq C\rho^n$.*

Proof. We first remark that there is an exhaustion of Ω by a monotonic sequence of relatively compact subdomains, one of which will contain the compact set K and the point a; since it is sufficient to prove the theorem with K replaced by the closure of this subregion, we may assume from the beginning that K is connected and contains the point a.

Next we note that for each $R > 0$ there exist positive numbers $r = r_R < R$ and $C = C_R$ with the following property: if h is any solution of $P(D)h = 0$ on a ball $B \equiv B(\xi, R)$ in $\mathbf{R}^N$, then there is a holomorphic function $\tilde{h}$ on the complex ball $\tilde{B} \equiv \tilde{B}(\xi, r)$ which agrees with h on the real ball $B(\xi, r)$ and satisfies $\|\tilde{h}\|_{\tilde{B}} \leq C\|h\|_B$ (see the proof of [1, Lemma 2]).

Now for each $\xi \in K$ we can find a radius $R(\xi)$ such that $B \equiv B(\xi, R(\xi)) \subset \Omega$, and we may apply the remark above to construct a holomorphic function $\tilde{f}_\xi$ on the ball $\tilde{B} \equiv \tilde{B}(\xi, r_{R(\xi)})$ which agrees with f on the real ball $B(\xi, r_{R(\xi)})$ and satisfies $\|\tilde{f}_\xi\|_{\tilde{B}} \leq C_{R(\xi)}\|f\|_B \leq C_{R(\xi)}$. We may now use the compactness of K to find a finite set $\mathcal{F} \subset K$ such that the smaller real balls $\{B(\xi, r_{R(\xi)})\}_{\xi \in \mathcal{F}}$ cover K. Then

$$\tilde{X} \equiv \bigcup_{\xi \in \mathcal{F}} \tilde{B}(\xi, r_{R(\xi)})$$

is a bounded domain in $\mathbf{C}^N$. Now let $\tilde{f}$ be the unique holomorphic function in $\tilde{X}$ whose restriction to $\tilde{B}(\xi, r_{R(\xi)})$ is $\tilde{f}_\xi$, for each $\xi \in \mathcal{F}$. Then $\|\tilde{f}\|_{\tilde{X}} \leq \tilde{C} \equiv \max\{C_{R(\xi)} : \xi \in \mathcal{F}\}$. Since $D^\alpha \tilde{f}(a) = 0$ for $|\alpha| < n$, we may apply Corollary 2.3 to the function $\tilde{f}/\tilde{C}$ to conclude that

$$|\tilde{f}(z)| = |f(x)| \leq \tilde{C}\rho^n \qquad \text{for } z = x \in K,$$

where $\rho = \sup_K[\exp g_{\tilde{X}}(\cdot, a)]$.

In Section 3 we will need the following Schwarz lemma for harmonic functions on unbounded domains, obtained from Theorem 2.4 by use of the Kelvin transform.

COROLLARY 2.5. *Let G be an unbounded domain in $\mathbf{R}^N$ such that the complement $\mathbf{R}^N \backslash G$ is a nonempty compact set disjoint from the closed subset L of $\mathbf{R}^N$. Then there exist constants $C > 0$ and $\rho \in (0, 1)$, depending only on G and L, with the following property. If u is a continuous function on $\bar{G}$ which is harmonic on G, and there exist $n > N - 2$, $A > 0$, and $R > 0$ such that*

$$|u(x)| \leq (A/|x|)^n \qquad \text{for } |x| > R,$$

then $|u(x)| \leq C\rho^n \max_{\partial G} |u|$ for all $x \in L$.

Proof. We may assume that G does not contain the origin. Fix a positive number r so that G contains $\{x : |x| \geq r\}$. Let $T(x) \equiv x/|x|^2$. Then the domain $\Omega = T(G) \cup \{0\} \subset \mathbf{R}^N$ contains the closed ball $\bar{B}(0, 1/r)$, and $K = T(L) \cup \{0\}$ is a compact subset of Ω. Let u satisfy the hypotheses of the theorem, and consider the harmonic function $f(x) = |x|^{2-N}u(T(x))$ in $\Omega \backslash \{0\}$. Then

$$|f(x)| \leq A^n |x|^{n-(N-2)} \qquad \text{for } 0 < |x| < 1/R,$$

so f can be extended to a harmonic function on all of Ω with $D^\alpha f(0) = 0$ for $|\alpha| < n - (N - 2)$. If $M = \max_{\partial G} |u|$, then $|f| \leq r^{N-2}M$ on $\partial \Omega$, so from the maximum principle this estimate also holds throughout Ω. Applying Theorem 2.4 to $(Mr^{N-2})^{-1}f$, we see that

$$|f(x)| \leq \tilde{C}M(r/\rho)^{N-2}\rho^n \qquad \text{for } x \in K,$$

where $\tilde{C} > 1$ and $\rho \in (0,1)$ are constants depending only on Ω and K. If d is the distance from the origin to L, it follows that for $x \in L$ we have

$$|u(x)| = \frac{1}{|x|^{N-2}}|f(x/|x|^2)|$$

$$\leq \frac{1}{|d|^{N-2}}|f(x/|x|^2)|$$

$$\leq \tilde{C}M(r/\rho d)^{N-2}\rho^n,$$

which proves the corollary with $C = \tilde{C}(r/\rho d)^{N-2}$.

3. Harmonic functions have the approximation property

THEOREM 3.1. *Let K be a compact subset of $\mathbf{R}^N$ such that $\mathbf{R}^N \backslash K$ is connected. Let Ω be an open neighborhood of K. Then there exists a constant $\rho < 1$, depending only on K and Ω, such that for any harmonic function f on Ω we have $\limsup_{n \to \infty} d_n(f, K)^{1/n} \leq \rho$.*

Proof. We give the proof for $N \geq 3$, and let $E(x) \equiv c_N|x|^{2-N}$ be the standard fundamental solution for the Laplace operator. We may assume that Ω is bounded. If n is fixed, it follows from the Hahn-Banach theorem and the Riesz representation theorem that there exists a signed Borel measure $\mu = \mu_n$ supported on K, with total variation 1, such that

$$\int p\,d\mu = 0 \qquad \text{for all } p \in \mathcal{H}_n, \tag{3}$$

and

$$\int f\,d\mu = d_n(f, K) \equiv d_n.$$

We next fix a function $\psi \in C_0^\infty(\Omega)$ which is identically equal to one on an open neighborhood D of K. We regard $F \equiv \psi f \in C_0^\infty(\Omega) \subset C_0^\infty(\mathbf{R}^N)$, and we have

$$d_n = \int_K f\,d\mu = \int F\,d\mu = \mu * F(0).$$

If we use the fact that

$$\mu * F = \mu * (F * \delta) = \mu * (F * \Delta E) = \Delta(\mu * F * E) = \tilde{\mu} * \Delta F,$$

where $\bar{\mu}(x) \equiv (\mu * E)(x) = c_N \int_K |x - y|^{2-N} d\mu(y)$, we obtain

$$d_n = \int_{\Omega \setminus D} \bar{\mu} \Delta F \, d\lambda. \tag{4}$$

The domain $\mathbf{R}^N \cup \{\infty\} \setminus K$ has an exhaustion by a monotonic sequence of relatively compact subdomains, and one of these subdomains must contain the compact set $\mathbf{R}^N \cup \{\infty\} \setminus D$; we define G to be this subdomain, with the point at infinity removed. Using the fact that the total variation of μ is 1, we see that we have

$$|\bar{\mu}(x)| \leq A \qquad \text{for } x \in \partial G, \tag{5}$$

where the bound A depends only on the distance between the sets K and G.

To estimate $\bar{\mu}$ at other points, we use a Taylor expansion for the fundamental solution (cf. [8], [3, Section 3]). We fix $x \neq 0$, and write the Taylor series for the function $y \to E(x - y)$:

$$E(x - y) \equiv \sum_{\ell=0}^{\infty} Q_\ell^{(x)}(y),$$

where

$$Q_\ell^{(x)}(y) = (-1)^\ell \sum_{|\alpha|=\ell} \frac{D^\alpha E(x) y^\alpha}{\alpha!} \in \mathcal{H}_\ell.$$

We next note that there exists a constant M, depending only on the dimension N, such that

$$|Q_\ell^{(x)}(y)| \leq |x|^{2-N-\ell} (M|y|)^\ell \sum_{|\alpha|=\ell} 1 \qquad \text{for } x \in \mathbf{R}^N \setminus \{0\} \text{ and } y \in \mathbf{R}^N. \tag{6}$$

In fact, since E is real-analytic on $\mathbf{R}^N \setminus \{0\}$, we may apply the Cauchy inequalities and the Heine-Borel theorem to see that there exists a constant M such that for each multi-index α we have

$$|D^\alpha E(x)| \leq \alpha! M^{|\alpha|} \qquad \text{if } |x| = 1.$$

Then (6) follows from the fact that for each multi-index α, the function $D^\alpha E$ is homogeneous of degree $2 - N - |\alpha|$ on $\mathbf{R}^N \setminus \{0\}$.

Now if $R = \sup_{y \in K} |y|$, then for $|x| > 2MR$ we have

$$|\tilde{\mu}(x)| = \left| \int_K E(x-y)\, d\mu(y) \right|$$

$$= \left| \int_K \sum_{\ell=n+1}^{\infty} Q_\ell^{(x)}(y)\, d\mu(y) \right|$$

$$\leq \sum_{\ell=n+1}^{\infty} \sup_{y \in K} |Q_\ell^{(x)}(y)|$$

$$\leq \frac{1}{(2MR)^{N-2}} \sum_{|\alpha| \geq n+1} \left(\frac{MR}{|x|} \right)^{|\alpha|}$$

$$\leq \frac{1}{(2MR)^{N-2}} \sum_{|\alpha| \geq n+1} \left(\frac{2MR}{|x|} \right)^{n+1} \left(\frac{1}{2} \right)^{|\alpha|}$$

$$\leq \frac{2^N}{(2MR)^{N-2}} \left(\frac{2MR}{|x|} \right)^{n+1},$$

where the first inequality follows from the fact that the total variation of μ is 1, the second inequality follows from (6), and the last inequality follows from the equation $\sum_{|\alpha| \geq 0} 1/2^{|\alpha|} = (\sum_{m=0}^{\infty} 1/2^m)^N = 2^N$. Using this estimate and (5), we may apply Corollary 2.5 to obtain constants $C > 0$ and $\rho \in (0,1)$ such that

$$|\tilde{\mu}(x)| \leq 2^N (2MR)^{2-N} AC\rho^{n+1}$$

for $x \in \mathbf{R}^N \backslash D$. From this estimate and (4) we see that

$$d_n \leq 2^N (2MR)^{2-N} AC\rho^{n+1} \cdot s \cdot \lambda(\Omega \backslash D),$$

where the number $s = \sup_{\Omega \backslash D} |\Delta F|$ depends on the function f but not on n. From this it follows that $\limsup_{n \to \infty} d_n^{1/n} \leq \rho$, which proves Theorem 3.1.

4. Functions having the approximation property are harmonic

We say that a compact set $K \subset \mathbf{R}^N$ is a *harmonic Bernstein-Walsh set* provided that for every number $b > 1$ there exist a constant $M = M_b > 0$ and an open neighborhood $U = U_b$ of K in $\mathbf{R}^N$ such that

$$\|p\|_U \leq M\, b^{\deg p} \|p\|_K$$

for all harmonic polynomials p on $\mathbf{R}^N$. This concept is due to Siciak [14], [15] (who used the terminology "condition (H)") and is closely related to pluripotential theory for functions of several complex variables [10]. Nguyen Thanh Van and B. Djebbar [11, Proposition 3] have proved that if $N = 2$ and K is a harmonic Bernstein-Walsh set in $\mathbf{R}^2$, then Theorem 1.1 holds when "holomorphic" is replaced by "harmonic"; their theorem implies our results in the case $N = 2$.

THEOREM 4.1. *Let K be a compact subset of $\mathbf{R}^N$ which is a harmonic Bernstein-Walsh set. If f is a bounded function on K such that*

$$\rho \equiv \limsup_{n \to \infty} d_n(f, K)^{1/n} < 1,$$

then f is harmonic on K.

Proof. We choose for each n a polynomial q_n in the finite dimensional space $\mathcal{H}_n$ so that $d_n(f, K) = \|f - q_n\|_K$, and note that by hypothesis we have $f \equiv q_0 + \sum_{n=0}^{\infty}[q_{n+1} - q_n]$ on K. If $b > 1$ is chosen such that $\rho b < 1$, then this series is uniformly Cauchy on U_b, and the sum must be harmonic there since it is a uniform limit of harmonic functions.

Theorems 3.1 and 4.1 together give the most general harmonic Bernstein theorem obtained by our methods: if K is a compact, harmonic Bernstein-Walsh subset of $\mathbf{R}^N$ such that $\mathbf{R}^N \backslash K$ is connected, then *a bounded function $f : K \to \mathbf{R}$ is harmonic on K if and only if (1) holds.*

To complete the proof of Theorem 1.2, we may appeal to a theorem of Plesniak [12] which states that under the hypotheses of Theorem 1.2, the set K must be a harmonic Bernstein-Walsh set. For completeness we include a direct proof of Plesniak's theorem. To do this we introduce for each nonempty compact set $K \subset \mathbf{C}^N$ the *extremal function*

$$u_K(z) \equiv \max \left\{ 0, \sup_p \left\{ \frac{1}{\deg p} \log |p(z)| \right\} \right\},$$

where the supremum is taken over the family of all nonconstant holomorphic polynomials p on $\mathbf{C}^N$ such that $\|p\|_K \leq 1$. The extremal function was introduced by Siciak [13] and is discussed in detail in [10], but in order to keep our presentation self-contained we will prove only some simple facts which suffice for the proof of Plesniak's theorem. From the definition we see that u_K is lower semicontinuous and nonnegative on $\mathbf{C}^N$, and $u_K \equiv 0$ on K. From the definition we also obtain the *monotonicity property*: if $K \subset K'$, then $u_{K'} \leq u_K$ on $\mathbf{C}^N$. The Green functions g_K used in the following remark were defined following Theorem 1.1.

REMARKS 4.2. (a) *Let* $K = K_1 \times \cdots \times K_N$, *where each* K_j *is a compact subset of* $\mathbf{C}$ *such that* $\mathbf{C} \backslash K_j$ *is connected and regular for the Dirichlet problem. Then*

$$u_K(z_1, \ldots, z_N) \leq \sum_{j=1}^{N} g_{K_j}(z_j). \tag{7}$$

In particular, u_K *is locally bounded on* $\mathbf{C}^N$, *and is continuous at each point of* K.

(b) *Let* $K \subset \mathbf{R}^N$ *be a compact set with nonempty interior,* $\text{int} K$, *in* $\mathbf{R}^N$. *If we regard* $K \subset \mathbf{R}^N = \mathbf{R}^N + i0 \subset \mathbf{C}^N$, *then* u_K *is locally bounded on* $\mathbf{C}^N$, *and is continuous at each point of* $\text{int} K$.

(c) *Let* K *be a compact subset of* $\mathbf{C}^N$ *such that* u_K *is continuous at each point of* K. *Then for each* $b > 1$ *there exists an open neighborhood* $V = V_b$ *of* K *in* $\mathbf{C}^N$ *such that* $\|p\|_V \leq b^{\deg p} \|p\|_K$ *for all holomorphic polynomials* p *on* $\mathbf{C}^N$.

We prove Remark 4.2 (a) by induction on N. If $N = 1$, and $p \equiv p(z)$ is a nonconstant polynomial satisfying $\|p\|_K \leq 1$, then $V \equiv \frac{1}{\deg p} \log |p| - g_K$ is a subharmonic function on $\mathbf{C} \cup \{\infty\} \backslash K$ which continuously assumes nonpositive values at each point of ∂K; thus $V \leq 0$ on $\mathbf{C} \cup \{\infty\} \backslash K$, and (7) follows. If $N \geq 2$, and $p \equiv p(z_1, \ldots, z_N)$ is a polynomial of degree $n \geq 1$ satisfying $\|p\|_K \leq 1$, we may freeze $z_N \in K_N$ and apply the induction hypothesis to obtain $|p(z_1, \ldots, z_N)| \leq \exp\left\{ n \sum_{j=1}^{N-1} g_{K_j}(z_j) \right\}$ for all $(z_1, \ldots, z_{N-1}) \in \mathbf{C}^{N-1}$; we then freeze $(z_1, \ldots, z_{N-1}) \in \mathbf{C}^{N-1}$ and apply the one-variable result to obtain (7).

Remark 4.2 (b) follows from applying 4.2 (a) and the monotonicity property of the extremal function to sets in $\text{int} K$ which are products of N nondegenerate intervals. To prove Remark 4.2 (c), note that by hypothesis each point $\xi \in K$ has an open neighborhood G_ξ in $\mathbf{C}^N$ on which $u_K < \log b$, so we may take $V_b = \bigcup_{\xi \in K} G_\xi$.

We should mention that Siciak has proved that equality holds in (7) if $\sum_{j=1}^{N} g_{K_j}(z_j)$ is replaced by $\max_{j=1}^{N} g_{K_j}(z_j)$; see [13, Section 8], [10]. We also mention that a strong converse of 4.2 (c) is known: if $K \subset \mathbf{C}^N$ is compact, and for each $b > 1$ there exist a constant $M = M_b > 0$ and an open neighborhood $V = V_b$ of K in $\mathbf{C}^N$ such that $\|p\|_V \leq M \, b^{\deg p} \|p\|_K$ for all holomorphic polynomials p in $\mathbf{C}^N$, then u_K is continuous on $\mathbf{C}^N$; see [19],[15],[10].

THEOREM 4.3 (Plesniak). *Let* K *be a compact subset of* $\mathbf{R}^N$. *such that every point of* ∂K *is analytically accessible from* $\text{int} K$. *Then* K *is a harmonic Bernstein-Walsh set.*

Proof. From Remark 4.2 (b) and the theory of plurisubharmonic functions it follows that the upper-semicontinuous regularization

$$u_K^*(z) \equiv \limsup_{\zeta \to z} u_K(\zeta)$$

is a non-negative plurisubharmonic function on $\mathbf{C}^N$ which is identically equal to zero on $\operatorname{int} K$.

We next prove that if $a \in \partial K$, then $u_K^*(a) = 0$, and hence u_K is continuous at a; in view of Remarks 4.2 (b) and (c), this will complete the proof of Theorem 4.3. By hypothesis there exists an open neighborhood G of $[-1, 1]$ in $\mathbf{C}$, and a holomorphic function $h : G \to \mathbf{C}^N$, such that (2) holds. Then the composition $v(z) \equiv u_K^*(h(z))$ is defined and subharmonic on an open neighborhood of the origin in $\mathbf{C}$, and has the value zero on all points of the positive real axis sufficiently near the origin. Since the positive real axis is not thin at the origin, we conclude that $v(0) = 0$. Thus $u_K^*(a) = 0$, which completes the proof.

We close the paper by recalling an example of Walsh [17, pages 515–516] which shows some limitations on the techniques used here. The interval $K = [-1, 1] \subset \mathbf{R}^2$ is not a harmonic Bernstein-Walsh set in $\mathbf{R}^2$, since the nonconstant harmonic polynomial $u(x, y) \equiv y$ vanishes on K. Nevertheless, it is true that *a bounded function $f : K \to \mathbf{R}$ is harmonic on K if and only if* (1) *holds*. In fact, a harmonic function on K must satisfy (1) by applying Theorem 3.1 above or the arguments of [11]. Conversely, let us now suppose that (1) holds for a function $f : K \to \mathbf{R}$, and prove that in fact f *must extend to a holomorphic function F on an open neighborhood of K*; then $\operatorname{Re} F$ will be an extension of f to a harmonic function on an open neighborhood of K. For the proof, we note that for each harmonic polynomial $p \equiv p(x, y)$ of degree n in $\mathbf{R}^2$, the function $p(\cdot, 0) : \mathbf{R} \to \mathbf{R}$ is a polynomial of degree $\leq n$ on $\mathbf{R}$ and hence extends to a holomorphic polynomial of degree $\leq n$ on $\mathbf{C}$. Using the notation of Theorem 1.1, we conclude that $D_n(f, K) \leq d_n(f, K)$ for each n. It then follows from (1) and Theorem 1.1 that f extends to a holomorphic function on an open neighborhood of K.

References

[1] D. H. Armitage, T. Bagby, and P. M. Gauthier, Note on the decay of elliptic equations, *Bull. London Math. Soc.* **17** (1985), pp. 554–556.

[2] D. H. Armitage and M. Goldstein, Better than uniform approximation on closed sets by harmonic functions with singularities, *Proc. London Math. Soc.* (4) **60** (1990), pp. 319–343.

[3] T. Bagby, Approximation in the mean by solutions of elliptic equations, *Trans. Amer. Math. Soc.* **281** (1984), pp. 761–784.

[4] M. S. Baouendi and C. Goulaouic, Approximation of analytic functions on compact sets and Bernstein's inequality, *Trans. Amer. Math. Soc.* **189** (1974), pp. 251–261.

[5] T. Bloom, On the convergence of multivariate Lagrange interpolants, *Constructive Approximation* **5** (1989), pp. 415–435.

[6] J.-P. Demailly, Mesures de Monge-Ampère et mesures plurisousharmoniques, *Math. Zeitschrift* **194** (1987), pp. 519–564.

[7] S. Dineen, *The Schwarz lemma*, Oxford Mathematical Monographs, Clarendon Press, Oxford, 1989.

[8] R. Harvey and J. C. Polking, A Laurent expansion for solutions to elliptic equations, *Trans. Amer. Math. Soc.* **180** (1973), pp. 407–413.

[9] M. Klimek, Extremal plurisubharmonic functions and invariant pseudodistances, *Bull. Soc. Math. France* **113** (1985), pp. 231–240.

[10] M. Klimek, *Pluripotential theory*, to appear.

[11] Nguyen Thanh Van and B. Djebbar, Propriétés asymptotiques d'une suite orthonormale de polynômes harmoniques, *Bull. Sc. math.* (5) **113** (1989), pp. 239–251.

[12] W. Plesniak, On some polynomial conditions of the type of Leja in $\mathbb{C}^n$, in *Analytic functions Kozubnik 1979, Proceedings*, Lecture Notes in Mathematics vol. 798, Springer-Verlag, pp. 384–391.

[13] J. Siciak, On some extremal functions and their applications in the theory of analytic functions of several complex variables, *Trans. Amer. Math. Soc.* **105** (1962), pp. 322–357.

[14] J. Siciak, Asymptotic behavior of harmonic polynomials bounded on a compact set, *Ann. Polon. Math.* **20** (1968), pp. 267–278.

[15] J. Siciak, Extremal plurisubharmonic functions in $\mathbb{C}^N$, *Ann. Polon. Math.* **39** (1981), pp. 175–211.

[16] J. L. Walsh, *Interpolation and approximation by rational functions in the complex domain*, Amer. Math. Soc. Coll. Publ. vol. 20, Third Edition, 1960.

[17] J. L. Walsh, The approximation of harmonic functions by harmonic polynomials and by harmonic rational functions, *Bull. Amer. Math. Soc.* **35** (1929), pp. 499–544.

[18] J. L. Walsh, Maximal convergence of sequences of harmonic polynomials, *Ann. of Math.* **38** (1937), pp. 321-364.

[19] V.P. Zaharjuta, Extremal plurisubharmonic functions, orthogonal polynomials and Bernstein-Walsh theorem for analytic functions of several complex variables, *Ann. Polon. Math.* **33** (1976), pp. 137-148.

Department of Mathematics   Department of Mathematics and Statistics
Swain Hall–East   University of Auckland
Indiana University   Private Bag
Bloomington, Indiana 47405 USA   Auckland, New Zealand

SPECTRAL THEORY OF NONLINEAR EQUATIONS AND N-WIDTHS OF SOBOLEV SPACES

A.P.Buslaev, V.M.Tikhomirov

1. Introduction

Let us consider the simplest differential equation of Sturm–Liouville type:

$$\ddot{x} + \lambda^2 x = 0, \quad x(0) = \dot{x}(1) = 0. \tag{1.1}$$

This equation is the Euler equation of the isoperimetrical problem in variational calculus

$$\int_0^1 x^2 dt \to \sup, \quad \int_0^1 \dot{x}^2 dt \le 1, \quad x(0) = 0$$

Written in another way it expresses the stationary po. its of Rayleigh ratio $(I = [0,1], \; \| \cdot \|_{L_p(I)} \iff \| \cdot \|_p)$

$$R'(x) = 0, \quad R(x) = R(x,1,2,2) = \|x\|_2 / \|\dot{x}\|_2. \tag{1.2}$$

There is a cour table number of solutions of (1.2)

$$x_n(t) = 2^{-1/2} \sin(n + 1/2)\pi t, \quad n \in \mathbf{Z}_+, \quad \|\dot{x}\|_2 = 1$$

and the corresponding eigenvalues (i.e. the spectral numbers) are $\lambda_n = (n + 1/2)\pi$. We denote the set $\{x_n\}_{n \in \mathbf{Z}_+}$ as $SP(1,2,2)$.

The study of Sturm–Liouville equations is one of the most important chapters of linear analysis. Equations similar to (1.1) appeared in memoirs of D.Bernulli, Euler, Lagrange and other classics of the 18-th century calculus. These equations have an enormous number of applications to classic and quantum mechanics, physics, engineering and classical analysis [1–6].

One of such applications was discovered by A.Kolmogorov in 1936, [6]. Let $W_2^1(\Gamma_0)$ be the Sobolev class of functions

$$W_2^1(\Gamma_0) = \{x \mid x \in AC(I), \; \|\dot{x}\|_2 \le 1, \; x(0) = 0\},$$

where $AC(I)$ is the space of absolutely continuous functions. Kolmogorov posed the problem to find the best n-dimensional subspace $\hat{L}_n$ that approximates a set C in a normalized space X. Thus he introduced the quantity

$$d_n(C, X) = \inf_{L_n \subset Lin_n(X)} \sup_{x \in C} \inf_{y \in L_n} \|x - y\| \tag{1.3}$$

where $Lin_n X$ is the set of all n-dimensional subspaces of X. It was called the Kolmogorov n-width. He proved that

$$d_n(W_2^1(\Gamma_0), L_2(I)) = \lambda_n^{-1} = ((n + 1/2)\pi)^{-1}. \tag{1.4}$$

This equality expresses a connection between the spectrum of (1.1) and the n-widths of Sobolev class $W_2^1(\Gamma_0)$ in L_2.

Let us now consider the generalized (p, q) – Rayleigh ratio

$$R(x, 1, p, q) = \|x\|_q / \|\dot{x}\|_p, \quad x(0) = 0, \quad 1 \le p, q \le \infty \tag{1.5}$$

and the problem of n-widths of the classes $W_p^1(\Gamma_0)$ in $L_q(I)$. Then the following question arises: is there a connection between the stationary points of Rayleigh's ratio and the n-widths of the pair $(W_p^1(\Gamma_0), L_q(I))$? This paper deals with such problems.

The stationary conditions for $R(x, 1, p, q)$ lead us to a nonlinear equations of Sturm–Liouville type

$$((\dot{x})_{(p)})' + \lambda^q(x)_{(q)} = 0, \quad x(0) = \dot{x}(1) = 0, \tag{1.5a}$$

where $a \in \mathbb{R} \implies (a)_{(p)} := |a|^{p-1} \operatorname{sgn} a$. It is more convenient to write these equations in their canonical form:

$$\dot{x} = (y)_{(p')}, \quad \dot{y} = -\lambda^q(x)_{(q)}, \quad x(0) = y(1) = 0, \quad (p')^{-1} + p^{-1} = 1.$$

Similar equations appear in many problems of physics, mechanics and mathematics. But the nonlinear case has many peculiarities compared to the linear one. We illustrate this idea in some simple examples.

Let us consider a discrete analog of the Rayleigh ratio (1.5)

$$R(x, A, p, q) = \|Ax\|_{l_q^m} / \|x\|_{l_p^m} \tag{1.6}$$

and two simple examples of the problem, when $m = 3$, $q = 2$, $p = 2$ or ∞,

$$Ax = A(x_1, x_2, x_3) = (x_1/a_1, x_2/a_2, x_3/a_3), \quad a_1 > a_2 > a_3.$$

For $p = q = 2$ there exist only six spectral vectors, i.e. the stationary points of $R(x, A, 2, 2)$, namely $\pm(a_1, 0, 0), \pm(0, a_2, 0), \pm(0, 0, a_3)$. In the case $p = \infty$, $q = 2$ we already obtain 27 spectral vectors: 8 vertices, the middle points of 12 edges and the middle points of 6 faces. In the case of the first example there exist three spectral numbers (a_1, a_2, a_3) coinciding with the n-widths of the set $Bl_2^3(a) = \{x \mid \sum_{i=1}^3 x_i/a_i^2 \le 1\}$ in the l_2^3 norm. In the second example there exist seven spectral numbers

$$(\sqrt{a_1^2 + a_2^2 + a_3^2}, \sqrt{a_1^2 + a_2^2}, \sqrt{a_1^2 + a_3^2}, \sqrt{a_2^2 + a_3^2}, a_1, a_2, a_3).$$

We also have

$$d_0(Bl_\infty^3(a), l_2^3) = \sqrt{a_1^2 + a_2^2 + a_3^2}, \quad d_1 = \sqrt{a_2^2 + a_3^2}, \quad d_2 = a_3.$$

In other words the nonlinear case is quite more complicated then the linear one.

2. Statement of problems and the main results

We start with the discrete case. Let $A = (a_{ij})_{1 \leq i,j \leq \infty}$ be a $m \times m$ matrix, $x \in R^m$, $(\| \cdot \|_p \stackrel{\text{def}}{=} \| \cdot \|_{l_p^m})$ and

$$R(x, A, p, q) = \|Ax\|_q / \|x\|_p, \quad 1 \leq p, q \leq \infty$$

be a generalized (p, q) – Rayleigh ratio connected with A.

A stationary (normalized) point of this ratio, i.e. x such that $R'(x) = 0$, $\|x\|_p = 1$, satisfies the equation

$$A^\top (Ax)_{(q)} = \lambda^q (x)_{(p)}, \quad \|x\|_p = 1. \tag{2.1}$$

We denote the set of spectral pairs (λ, x) satisfying (2.1) by $SP(A, p, q)$.

In the continuous case we consider a kernel $K(\cdot, \cdot) : I^2 \to \mathbb{R}$, an integral operator defined by $K(\cdot, \cdot)$

$$Ku(t) = \int_0^1 K(t, \tau)u(\tau)d\tau,$$

and the corresponding Rayleigh ratio

$$R(u) := R(u, K, p, q) = \|Ku\|_q / \|u\|_p.$$

The stationary (normalized) points of this ratio, i.e. $u \in L_p(I)$ such that $R'(u) = 0$, $\|u\|_p = 1$, satisfy the equations

$$K^*(x)_{(q)} = \lambda^q (u)_{(p)}, \quad \|u\|_p = 1,$$

$$x(t) := Ku(t), \quad K^*v(t) = \int_0^1 K(\tau, t)v(\tau)d\tau. \tag{2.2}$$

The set of such functions is denoted by $SP(K, p, q)$. Put $W_p^K \stackrel{\text{def}}{=} \{x(\cdot) \mid x(\cdot) = Ku(\cdot), \|u\|_p \leq 1\}$.

For $p = q = 2$, $K(t, \tau) = (t - \tau)_+^0 = \{1, \text{ if } \tau \in [0, t] \text{ and } 0, \text{ if } \tau \notin [0, t]\}$ and we obtain (1.1).

The spectral theory of nonlinear equations of types (2.1) or (2.2) has an interesting application to problems of approximation. A question of comparing different methods of approximation leads to the notion of n-widths, which characterizes the optimal methods of approximation. In this connection we recall two definitions.

Let X be a normed space with the unit ball $BX = \{x \mid \|x\| \leq 1\}$ and C be a convex centrally symmetric subset of X. The Kolmogorov n-width of C in X is

$$d_n(C, X) = \inf_{L_n \subset Lin_n(X)} \inf_\varepsilon (\varepsilon BX + L_n \supset C) = \inf_{L_n \subset Lin_n(X)} \sup_{x \in C} \inf_{y \in L_n} \|x - y\|.$$

The Bernstein n-width is defined by

$$b_n(C,X) = \sup_{L_{n+1} \subset Lin_n(X)} \sup_\varepsilon (\varepsilon BX \cap L_{n+1} \subset C).$$

The Kolmogorov n-width characterizes the best approximative possibilities of n-dimensional subspaces, while the Bernstein n-widths characterize the best n-dimensional subspaces from the point of view of Bernstein's inequality.

Before stating the main result we introduce some more notation. For $x = (x_1,\ldots,x_m) \in R^m \backslash \{0\}$ we denote by $P_-(x)$ the number of sign changes in $\{x_1,\ldots,x_m\}$ with zero terms discarded. We denote by P_+ the maximum number of sign changes in $\{x_1,\ldots,x_m\}$ where zero terms have arbitrarily assigned values 1 or -1.

If $P_- = P_+$ we write P. If x is a function on I, then $P(x)$ denotes the number of sign changes of x, i.e. $P(x) = \max\{P(x(t_1),\ldots,x(t_n)), \ 0 \le t_1 \cdots < t_m \le 1\}$. We put

$$\text{Null } x(\cdot) = \{t \in I \mid x(t) = 0\},$$
$$SP_n(A,p,q) = \{x \in SP(A,p,q) \mid P(x) = n\},$$
$$SP_n(K,p,q) = \{x \in SP(K,p,q) \mid \text{Card Null } x|_{(0,1)} = 1\}.$$

THEOREM 1 ([16,17,19]). *Let* $1 \le p,q \le \infty$, $m \in N$, $A = (a_{i,j})$, $1 \le i,j \le m$, $\det A \ne 0$, $K(\cdot,\cdot) \in C^\infty(I^2)$ *and*

$$P_+(Ax) \le P_-(x), \ P_+(A^\mathsf{T}x) \le P_-(x) \ \forall x \in \mathbb{R}^n \backslash \{0\}, \qquad (2.3)$$

$$\begin{aligned}\text{Card Null}(Ku) \le Pu \quad &\forall u \in L_p(I), P(u) < \infty,\\ \text{Card Null}(K^*v) \le Pv \quad &\forall v \in L_p(I), P(v) < \infty.\end{aligned} \qquad (2.4)$$

Then

(α) $SP_n(A,p,q) \ne \varnothing, \ SP_n(K,p,q) \ne \varnothing$;
(β) *if* $1 \le q \le p \le \infty$ *then*

$$d_n(ABl_p^m, l_q^m) = \min\{\|Ax\|_q \mid x \in SP_n(A,p,q)\},$$
$$d_n(W_p^K, L_q(I)) = \min\{\|x\|_q \mid x \in SP_n(K,p,q)\};$$

(γ) *if* $1 \le p \le q \le \infty$ *then*

$$b_n(ABl_p^m, l_q^m) = \max\{\|Ax\|_q \mid x \in SP_n(A,p,q)\},$$
$$b_n(W_p^K, L_q(I)) = \max\{\|x\|_q \mid x \in SP_n(K,p,q)\};$$

3. Proof of the main Theorem

3.1 The first step: the spectrum is not empty. We consider the discrete case first.

Constructions of an iterative sequence. Let $\xi \in Bl_p^m$,

$$
\begin{aligned}
z^{(0)} &= \xi, \ z^{(s)} = z^{(s)}(\xi) = \nu_s(A^T(AZ^{(s-1)})_{(q)})_{(p')} \\
&\Longleftrightarrow A^T(AZ^{(s-1)})_{(q)} = \mu_s^q(z^{(s)})_{(p)}, \ \|z^{(s)}\|_p = 1 \\
&(|A| \neq 0 \Longrightarrow A^T((AZ^{(s-1)})_{(q)})_{(p')} \neq 0 \\
&\Longrightarrow \nu_s = \|(A^T((AZ^{(s-1)})_{(q)})_{(p')}\|_p^{-1}
\end{aligned}
\tag{3.1}
$$

LEMMA 1. μ_s and $\|Az(s)\|_q$ are monotone sequences.

Proof.

$$
\begin{aligned}
\|Az^{(s-1)}\|_q^q &\overset{\text{Id}}{=} \langle A^T(Az^{(s-1)})_{(q)}, \ z^{(s-1)} \rangle \overset{(3.1)}{=} \mu_s \langle (z_{(p)}^{(s)}, \ z^{(s-1)} \rangle \\
&\overset{\text{(Hölder)}}{\leq} \mu_s^q \|z^{(s)}\|_p^{(p-1)} \|z^{(s-1)}\|_p = (\text{NC-normalization condition}) \\
&= \mu_s^q = \mu_s^q \langle (z^{(s)})_{(p)}, \ z^{(s)} \rangle \overset{(3.1)}{=} \langle (Az^{(s-1)})_{(q)}, \ Az^{(s)} \rangle \\
&\overset{\text{(Hölder)}}{\leq} \|Az^{(s-1)}\|_q^{q-1} \|Az^{(s)}\|_q \Longrightarrow \|Az_q^{(s-1)} \leq \mu_s \leq \|Az^{(s)}\|_q
\end{aligned}
\tag{3.2}
$$

We now choose a subsequence $z^{s_k} (\|z^{(s)}\|_p = 1)$ such that $z^{(s_k)} \to \hat{z}$, $\|\hat{z}\|_p = 1$. Let $\mu_{s_k} \nearrow \mu$. Then

$$
\begin{aligned}
\mu_s^k &\overset{(3.2)}{\leq} \|Az^{(s_k)}\|_q^q \leq \mu_{s_k+1}^q \langle (z^{(s_k+1)})_{(p)}, \ z^{(s_k)} \rangle \leq \mu_{s_k+1}^q \\
&\Longrightarrow (\mu_{s_k} \longrightarrow \hat{\mu}, \ \mu_{s_k+1} \longrightarrow \hat{\mu}), \ \langle (z^{(s_k+1)})_{(p)}, \ z^{(s_k)} \rangle \to 1 \\
&\overset{\text{(Hölder)}}{\Longrightarrow} z^{(s_k+1)} \to \hat{z}, \ A^T(Az^{(s_k)})_{(q)} = \mu_{s_k+1}^q(z^{(s_k+1)})_{(p)} \\
&\Longrightarrow A^T(A\hat{z})_{(q)} = \hat{\mu}^q(\hat{z})_{(p)} \overset{\text{def}}{\Longrightarrow} \hat{z} \in SP(A,p,q).
\end{aligned}
\tag{3.3}
$$

Let $O^n = \partial Bl_1^{n+1} = \{y \in \mathbb{R}^{n+1} \mid \sum_{i=1}^{n+1} |y_i| = 1\}$ be the boundary of the $(n+1)$-dimensional octahedron. We construct the following function $u(\cdot,\cdot): I \times O^n \longrightarrow \mathbb{R}$,

$$
u(t,y) := \{\operatorname{sgn} y_k, \ t \in \Delta_k := \left[\sum_{i=0}^{k-1} |y_i|, \ \sum_{i=1}^{k} |y_i|\right], \ y_0 = 0\}.
\tag{3.4}
$$

Let $\xi = \xi(y)$ be a vector

$$\xi = (\xi_1, \ldots, \xi_n), \ \xi_j := m \int_{\delta_j} u(t, y) dt, \ \delta_j := \left[\frac{j-1}{m}, \frac{j}{m} \right], \ 1 \le j \le m.$$

We proved that the sequence

$$A^\top (A z^{(s)}(y))_{(q)} = \mu_{s+1}^q(y)(z^{(s+1)}(y))_{(p)}, \ (\|z^{(s+1)}(y)\|)_{(p)} = 1, \ z^{(0)}(y) = \xi(y)$$

has a subsequence $\{z^{(s_k)}(y)\}_{k \in \mathbb{N}}$ such that

$$z^{(s_k)}(y) \longrightarrow \hat{z}(y) \in SP(A, p, q) \Longrightarrow P_-(z^{(s)}(y)) \overset{\text{def}}{=} P_+(A^\top(A z^{(s-1)}(y))_{(q)})$$
$$\overset{\text{cond}}{\le} P_-((A(z^{(s-1)}))_{(q)}) \overset{\text{def}}{\le} P_+((A(z^{(s-1)}(y)))_{(q)}) \le P_-((z^{(s-1)})_{(q)})$$
$$= P_-(z^{(s-1)}(y)) \le \cdots \le P(z^{(0)}(y)) \le n.$$

Let $Y(n, s)$ be the set $\{y \in O^n \mid P_+(z^{(s)}(y)) \ge n\}$.

LEMMA 2. $Y(n, s) \ne \varnothing$ for every $s, n \in \mathbb{Z}_+$.

Proof. Let us denote $F_s(y) = (z_1^{(s)}(y), \ldots, z_n^{(s)}(y))$. Then $F_s \in C(O^n, R^n)$ and $F_s(-y) = -F_s(y)$. By Borsuk's theorem there exists a vector $\bar{y}$ such that $F_s(\bar{y}) = 0$. By definition $P_+(z^{(s)}(\bar{y})) \ge n \Longrightarrow z^{(s)}(\bar{y}) \in Y(n, s)$

The set $Y(n, s)$ is closed and $P_+(z^{(s)}(y)) \le P_-(z^{(s-1)}(y))$ implies that $Y(n, 0) \supset Y(n, 1) \supset \ldots$. Hence $\hat{y} \in \cap_{s \in \mathbb{Z}_+} Y(n, s)$.

Let $z^{(s)}(\hat{y}) \to \hat{z}$. Since $P_+(z^{(s)}(\hat{y})) \ge n$ we have $P_+(\hat{z}) \ge n$ and therefore $SP_n(A, p, q) \ne \varnothing$.

The fact that $SP(K, p, q) \ne \varnothing$ can be proved in a similar way. We fix a function $u(\cdot, y)$ and construct an iterative sequence

$$u_0(\cdot, y) = u(\cdot, y), \ u_s(\cdot, y) = \nu_s(K^*(K u_s(\cdot, y))_{(q)})_{(p')}, \ \|u_s(\cdot, y)\|_p = 1. \quad (3.5)$$

Next we can prove statements analogues to Lemma 1 and Lemma 2 and finally obtain that $SP(K, p, q) \ne \varnothing$.

3.2 The second step is an estimate d_n from below.

LEMMA 3. *The following equality holds*

$$\lim_{s \to \infty} \min_{y \in O^n} \|A z^{(s)}(y)\|_q = \min_{y \in O_n} \lim_{s \to \infty} \|A z^{(s)}(y)\|_q (:= h). \quad (3.6)$$

Proof. Let $H(s, \varepsilon) = \{y \in O^n \mid \|A z^{(s)}(y)\|_q \le h - \varepsilon\}$. It is obvious that $H(s, \varepsilon)$ is a closed subset of O^n. It follows from Lemma 1 that the sequence

$\|Az^{(s)}(y)\|_q$ is monotonic, and therefore $H(0,\varepsilon) \supset H(1,\varepsilon) \supset \ldots$. Then either there exists s_0 such that $H(s,\varepsilon) \neq \varnothing$ for every $s \geq s_0$ and therefore

$$\|Az^{(s)}(y)\|_q > h - \varepsilon, s \geq s_0,$$
$$y \in O^n \Rightarrow \lim_{s \to \infty} \min_{y \in O^n} \|Az^{(s)}(y)\|_q \geq h - \varepsilon, \tag{3.7}$$

or $H(s,\varepsilon) \neq \varnothing$ for every s and there exists $\tilde{y} \in H(s,\varepsilon)$ such that $h := \min_{y \in O_n} \lim_{s \to \infty} \|Az^{(s)}(y)\|_q \leq \lim_{s \to \infty} \|Az^{(s)}(\tilde{y})\|_q \leq h - \varepsilon$ which is impossible. It follows that (3.7) holds for all ε.

Now let $1 < q < \infty$ and L be a subspace of $L_q(I)$ of dimension n. We consider the metric projection $\Phi_s(y) = Pr(Az^{(s)}(y), L, l_q^m)$ in l_q^m onto L. Clearly

$$\Phi_s(y) \in C(O^n, L), \ \dim L = n, \ \Phi_s(-y) = -\Phi_s(y).$$

By Borsuk theorem there exists a vector $\hat{y}$ such that $\Phi_s(\hat{y}) = 0$ which yields $d(Az^{(s)}(\hat{y}), L, l_q^m) = \|z^{(s)}(\hat{y})\|_q^!$. Therefore

$$d_n(ABl_p^m, l_q^m) \geq \lim_{s \to \infty} \min_{y \in O^n} \|Az^{(s)}(y)\|_q \overset{L3}{=} \min_{y \in O_n} \lim_{s \to \infty} \|Az^{(s)}(y)\|_q$$

$$= \min_{y \in O^n} \{\|Az^{(s)}(y)\|_q\} \overset{def}{=} \min\{\|Ax\|_q, x \in SP_n(A, p, q)\}.$$

3.3 The final step is to estimate d_n from above. We consider the continuous case. The discrete case is considered similarly.

Let $\bar{x} = K\bar{u} \in SP_n(K, p, q), \{\tau_i\}_{i=1}^n \in$ Null $\bar{u}(\cdot)$,

$$T = \begin{pmatrix} t_1 & \cdots & t_n \\ \tau_1 & \cdots & \tau_n \end{pmatrix}, \quad \bar{L}_n = \lim\{K(\cdot, \tau_j)\}_{j=1}^n.$$

It is easy to prove that a function $x \in W_p^K$ can be interpolated by a function $y \in L_n$ at the points $\{t_i\}_{i=1}^n$ and to obtain the formulae

$$x(t) - y(t) = \int_0^1 K_T(t, \tau) u(\tau) d\tau,$$

$$K_T(t, \tau) = K\begin{pmatrix} t_1 & \cdots & t_n & t \\ \tau_1 & \cdots & \tau_n & \tau \end{pmatrix} / K\begin{pmatrix} t_1 & \cdots & t_n \\ \tau_1 & \cdots & \tau_n \end{pmatrix}, \tag{3.8}$$

$$K\begin{pmatrix} t_1 & \cdots & t_n \\ \tau_1 & \cdots & \tau_n \end{pmatrix} = \det\left(K(t_i, \tau_j)\right)_{ij=1}^n$$

By definition $K(t, \tau_i) = 0$. Let us consider the problem

$$\|K_T u\| \longrightarrow \sup, \quad \|u\|_p \leq 1.$$

It is clear that this problem has a solution which we denote by $\hat{x} = K_T\hat{u}$. This solution satisfies the equation

$$\hat{\lambda}^q \int_0^1 (\hat{x})_{(q)}(t)K(t,\tau)dt = (\hat{u})_{(p)}(\tau), \ \|\hat{u}\|_p = 1.$$

If $x = K_T u$, then $|x(t)| \leq \int_0^1 |K(t,\tau)| u(\tau) d\tau$ for every $t \in I$, and $|x(t)| = \int_0^1 |K(t,\tau)| u(\tau) d\tau$ if and only if $K_T(t,\tau)u(\tau) \geq 0 \ \forall \tau \in I \implies P(\hat{u}) \leq n$ $(P(\bar{u}) = n) \implies \text{Card Null} \, \hat{x} \leq n \ (\text{Card Null} \, \bar{x} = n) \implies \text{Card Null} \, \hat{x} = n = \text{Card Null} \, \bar{x}, \text{sgn} \, \hat{u} = \text{sgn} \, \bar{u}.$ Then $\bar{\lambda} = R(\bar{x}) = \|\bar{x}\|_p^{-1}$, $f(t) = |\bar{x}(t)|^q \bar{\lambda}^q$. Below $(*)$ and $(**)$ refer to Iensen's and Iessen's inequalities. We have

$$\|K_T\bar{u}\|_q \overset{\text{def}}{\leq} \|K_T\hat{u}\|_q \overset{\text{def}}{=} \bar{\lambda}^{-1} \left(\int_0^1 f(t)|\hat{x}(t)/\bar{x}(t)|^q dt \right)^{1/q}$$

$$\overset{(*),p\geq q}{\leq} \bar{\lambda}^{-1} \left(\int_0^1 f(t)|\hat{x}(t)/\bar{x}(t)|^p dt \right)^{1/p}$$

$$= \bar{\lambda}^{-1} \left(\int_0^1 f(t) \left| \int_0^1 \frac{K_T(t,\tau)\hat{u}(\tau)}{\bar{x}(t)} d\tau \right|^p dt \right)^{1/p}$$

$$\overset{\text{Id}}{=} \bar{\lambda}^{-1} \left(\int_0^1 f(t) \left| \int_0^1 \frac{K_T(t,\tau)\bar{u}(\tau)\frac{\hat{u}(\tau)}{\bar{u}(\tau)} d\tau}{x(t)} \right|^p dt \right)^{1/p}$$

$$\overset{(**)}{\leq} \bar{\lambda}^{-1} \left(\int_0^1 \int_0^1 f(t) \frac{K_T(t,\tau)}{\bar{x}(t)} \left(\frac{\hat{u}(\tau)}{\bar{u}(\tau)} \right)^p \bar{u}(\tau) dt d\tau \right)^{1/p}$$

$$= \bar{\lambda}^{-1} \left(\int_0^1 \left(\frac{\hat{u}(\tau)}{\bar{u}(\tau)} \right)^p \bar{u}(\tau) \int_0^1 \bar{\lambda}^q(\bar{x})_{(q)} K_T(t,\tau) dt d\tau \right)^{1/p}$$

$$\overset{\text{def}}{=} \bar{\lambda}^{-1} \left(\int_0^1 \left(\frac{\hat{u}(\tau)}{\bar{u}(\tau)} \right)^p \bar{u}(\tau)(u(\tau))_{(p)} d\tau \right)^{1/p} = \bar{\lambda}^{-1}\|\hat{u}\|_p = \bar{\lambda}^{-1}.$$

4. Further results

Let $w_r, w_0 : I \longrightarrow R$ be continuous positive weight functions.

$$\|x\|_{q(w_0)} = \|x(\cdot)\|_{L_{q(w_0)}(I)} = \begin{cases} \left(\int_0^1 |x(t)|^q w_0(t) dt \right)^{(1/q)}, & 1 \leq q < \infty \\ \underset{t \in I}{\text{vrai sup}} \, |x(t)w_0|, & p = \infty, \end{cases}$$

$$W_{(pw_r)}^r(\Gamma) = \{x : I \to R \mid x^{(r-1)} \in AC(I), \|x^{(r)}\|_{p(w_r^{(-1)})} \leq 1, \ x|_{\partial I} \in \Gamma\}.$$

The stationary points of generalized Rayleigh ratio

$$R(x) = R(x, r, p, q, \Gamma, w_0, w_r) = \|x\|_{q(w_0)} / \|x^{(r)}\|_{p(w_r^{(-1)})}, \quad x|_{\partial I} \in \Gamma \quad (4.1)$$

are the solutions to a boundary problem for differential equations

$$(-1)^{r+1}((x_{(p)}^{(r)}/w_r)^{(r)} + \lambda^q(x)_{(q)}w_0 = 0,$$
$$x|_{\partial I} \in \Gamma, \quad ((x_{(p)}^{(r)}/w_r)|_{\partial I} \in \Gamma^\mathsf{T}, \quad (4.2)$$

where Γ^T is a transversal boundary condition. The set of spectral functions for equation (4.2) with n zeroes in $(0,1)$ is denoted by $SP(r, p, q, \Gamma, w_0, w_r)$. We write

$$x|_{\partial I} \notin \Gamma_0, \quad \text{if} \quad x^{(i)}(0) = (0), \ 0 \le i \le r - 1,$$
$$x|_{\partial I} \in \Gamma_0^\mathsf{T}, \quad \text{if} \quad x^{(i)}(1) = 0, \ 0 \le i \le r - 1,$$
$$x|_{\partial I} \in \overset{\circ}{\Gamma}, \quad \text{if} \quad x^{(i)}(0) = x^{(i)}(1) = 0, 0 \le i \le r - 1,$$
$$x|_{\partial I} \in \Gamma_a, \quad \text{if} \quad x^{(i)}((1 - (-1)^i)/2) = 0, \ 0 \le i \le r - 1,$$
$$x|_{\partial I} \in \Gamma_\varnothing, \quad \text{if} \quad x(\cdot) \text{ has no boundary conditions on } I,$$
$$x|_{\partial I} \in \tilde{\Gamma}, \quad \text{if} \quad x^{(i)}(0) = x^{(i)}(1), \ 0 \le i \le r - 1.$$

THEOREM 2 [15,16]. *Let* $r \in N$, $1 < p, q < \infty$, $\|w_0\|_{s_0} \le 1$, $\|w_r\|_{s_r} \le 1$, $s_0 \ge 1$, $s_r \ge (r-1)^{-1}$, $\Gamma \in \{\Gamma_0, \Gamma_0^\mathsf{T}, \overset{\circ}{\Gamma}, \Gamma_a, \Gamma_D, \tilde{\Gamma}\}$. *If* $n \ge n(r, \Gamma) = \{0, \ \Gamma = \Gamma_0 \vee \Gamma_0^\mathsf{T} \vee \overset{\circ}{\Gamma} \vee 1_a; \ r, \Gamma = \Gamma_\varnothing\}$ *then*
a) $SP_n(r, p, q, \Gamma, w_0, w_r) \ne \varnothing$, $\Gamma \ne \tilde{\Gamma}$; $SP_{2n}(r, p, q, \tilde{\Gamma}, w_0, w_r) \ne \varnothing$;
b) *if for every* t $w_0(t) > c_0 > 0$, $w_r(t) > c_r > 0$ *and* $\|w_i\|_{C(i)} \le D$, $i = 0 \vee r$
then for every $n \in \mathbb{Z}_+$ $\mathrm{Card}\, SP_n(r, p, q, \Gamma, w_0, w_r) < \infty$ *and unique if*

$\alpha)$ $1 < p = q < \infty$, or $\beta)$ $p > q$, $\Gamma = \Gamma_0 \vee \Gamma_a \vee \overset{\circ}{\Gamma} \vee \Gamma_0^\mathsf{T}$ and $n = 0$, or $\gamma)$ $1 < q < p < \infty$, $r = 1$, $\Gamma = \Gamma_0 \vee \Gamma_1 (= \Gamma_a \vee \Gamma_a^\mathsf{T})$, $w_0, w_1 \in C^1(I)$ and $w_1^{p'}(t)/w_0^p(t) = e^{a \int (w_0 / \int w_0 dt) dt}$ or $\delta)$ $1 < q < p < \infty$, $\Gamma = \Gamma_0$ and $(w_0(t) = \frac{d_1}{b} e^{c_0 t/b}$, $w_r^{-1}(t) = \frac{d_2}{b} e^{c_r t/b})$, or $w_0(t) = d_1(at + b)^{\frac{c_0}{a} - 1}$, $w_r^{-1}(t) = d_r(at + b)^{\frac{c_r}{a} - 1}$, where $c, d_1, d_2, a, b, c_0, c_r$ are some constants.

THEOREM 3 [14,16]. *Let* $r, n \in N$, $1 < q \le p < \infty$, $w_0 = w_r = 1$. *Then*

$$x_{2n} \in SP_{2n}(r, p, q, \tilde{\Gamma}, 1, 1) \implies x_{2n}(t) = z(nt),$$
$$z(t) = \{x_0(4t), \ 0 \le t \le 1/4; \ x_0(2 - 4t), \ 1/4 \le t \le 1/2;$$
$$-x_0(4t - 2), \ 1/2 \le i \le 3/4, \ -x_0(4 - 4t), \ 3/4 \le t \le 1\},$$

where $(\lambda_0, x_0) \in SP_0(r, p, q, \Gamma_a, 1, 1)$, $\mathrm{Card}\, SP_0(r, p, q, \Gamma_a, 1, 1) = 1$, $\lambda_0 = \lambda(r, p, q)$.

THEOREM 4 [18]. Let $r \in N$, $\Gamma \in \{\Gamma_0, \mathring{\Gamma}, \Gamma_\varnothing, \Gamma_a, \tilde{\Gamma}, \Gamma_0^\top, \Gamma_a^\top\}$, $1 \le q \le p \le \infty$, w_0 and w_r satisfy the conditions of the Theorem 1. Then

$$\lim_{n \to \infty} (n^r d_n(\mathring{W}^r_{p(w_r)}(\Gamma), L_{q(w_0)}) = \lambda_{rpq}^{-1} \left(\int_0^1 (w_0^{1/q} w_r^{1/p})^\mu dt \right)^{1/\mu},$$

where $\mu^{-1} = (r + q^{-1} - p^{-1})$.

The problem

$$\lambda_n(r, p, q, \Gamma, w_0, w_r) \to extr, \ \|w_i\|_{s_i} = 1, \ i = 0 \vee r$$

generalizes the Lagrange problem on the form of column optimizing the shapes of stands so that for a given volume the value of the limit load will be as large as possible.

THEOREM 5 [15]. Let $n \in Z_+$, $r \in N$, $\Gamma = \Gamma_0$, $1 < p = q < \infty$, w_0 and w_r satisfy the conditions of the Theorem 1. Then for $1 \le s_0 \le \infty$

$$\sup_{\|w_0\|_{s_0} = 1} \lambda_n^{-1}(r, p, p, \Gamma_0, w_0, w_r) = \sup\{\|x_n\|_{ps_0'}, x_n \in SP_n(r, p, ps_0', \Gamma_0, 1, w_r)\}.$$

For $s_r \ge (p-1)^{-1}$

$$\sup_{\|w_r\|_{s_r} = 1} \lambda_n^{-1}(r, p, p, \Gamma_0, w_0, w_r)$$
$$= \sup\{\|x_n\|_{q(w_0)}, x_n \in SP_n(r, p(\cdot s_r)', q, \Gamma_0, w_0, 1)\}.$$

5. Concluding remarks

The theorem presented in section 2 has a long history [10]. In his paper devoted to the problem of n-widths [6,10] A.Kolmogorov calculated n-widths of Sobolev classes $W_2^r(S)$ in $L_2(S)$, $S = I, \mathbb{T}$. His proof was based on the orthogonal property of functions $x_n \in SP(r, 2, 2)$.

A different approach to this problem was given by V.Tichomirov in his papers of 1960–69 [7–10]. For example consider the problem on $d_n(W_p^1(\Gamma_0))$ in $L_q(I)$, $p \ge q$ we started with in section 1. Equation (1.5a) can be integrated in quadratures. We obtain solutions x_n of (1.5a) with n zeros. Next we observe that an optimal method of approximation of $W_p^1(\Gamma_0)$ in $L_q(I)$ is interpolation of $x \in W_p^1(\Gamma_0)$ at zeros of x_n by a picewise constant function y with gaps at the zeroes of $\dot{x}_n$ [9,11].

We saw that the proof of the Theorem 1 [16,17,19] consists of three parts and is based on the following ideas. The initial point of the second step (the estimate from below) is due to Tichomirov (1960, [7], $r = 1, p = q$). It was generalized by Makovoz (1972, [11], $r = 1, p \geq q$). The method for proving the existence of the spectrum, obtaining lower bounds for Kolmogorov n-widths and so upper bounds for Bernstein n-widths was developed by Buslaev (1989, [15]).

The method of approximation appeared for the first time in Tichomirov's paper (1969, [8]). The first discussion of the discrete case and an application of methods of ETP matrix to these problems appeared for the first time in papers of American mathematicians Micchelli, Pinkus and others – the school of Carlin [13].

The case of $p = q$ was proved by Pinkus (1985, [12]). The case $p > q$ [17] and the results of section 4 were proved by Buslaev.

References

[1] Euler, *Collected Works*, Moscow, 1934. (In Russian)

[2] Lagrange J.L., *Eurves*, Gauthier–Villars, Paris, 1968.

[3] Rayleigh J., *The Theorie of the Sound*, New York, 1945.

[4] Courant R. and Hilbert D., *Methoden der mathematishen physik*, Springer, Berlin, 1934.

[5] Dunford N. and Schwartz J, *Linear Operators. Part 2. Spectral Theory*, Interscience publishers, New York, London, 1963.

[6] Kolmogorov A. N., *Uber die beste Annaherung von Functionen einegeben Funktionen-klasse*, Ann. of Math. **37** (1936), pp. 107–110.

[7] Tikhomirov V.M., *Diameters of sets in function spaces and the theory of best approximations*, Uspechi Math. Nauk **15** (1960), pp. 81–120. (In Russian)

[8] Tikhomirov V.M., *Best method of approximation and interpolation of difference functions in the space* $C[-1, 1]$, Math. Sbornik **80(122)** (1969), pp. 290–304. (In Russian)

[9] Tikhomirov V.M., *Some Problems in Theory of Approximation*, MGU, Moscow, 1976. (In Russian)

[10] Tikhomirov V.M., *A.N. Kolmogorov and Approximation Theory*, Uspechi Math. Nauk **44** (1989), pp. 83–122. (In Russian)

[11] Makovoz Yu.I., *On a method for estimation from below of dimeters of sets in Banach spaces*, Math. Sbornik **87** (1972), pp. 136–142. (In Russian)

[12] Pinkus A., *Some extremal problems for strictly torally positive matrices*, Linear Alg. and Appl. **64** (1985), pp. 141–156.

[13] Pinkus A., Constructive approximation **1** (1985), pp. 15–62.

[14] Buslaev A.P. and Tikhomirov V.M., *Some problems of nonlinear analysis*, Dokl. Akad. Nauk. SSSR **283** (1985), pp. 13–18. (In Russian)

[15] Buslaev A.P., *Extremal problems of approximation theory and nonlinear oscillations*, Dokl. Akad. Nauk. SSSR **305** (1989), pp. 379–384. (In Russian)

[16] Buslaev A.P. and Tikhomirov V.M., *Spectra of nonlinear differential equations and widths of Sobolev spaces*, Math.Sbornik **181** (1990), pp. 1587–1606. (In Russian)

[17] Buslaev A.P., *On variational description of the spectrum of totally positive matrices and extremal problems of the approximation theory*, Mat.Zametki **49** (1990), pp. 39–46. (In Russian)

[18] Buslaev A.P., *On asymptotics of widths and spectra of nonlinear differential equations*, Algebra i Analiz **3** (1991). (In Russian)

[19] Buslaev A.P., *On Bernstein–Nikolsky inequality and widths of Sobolev classes*, Dokl. Akad. Nauk. (1992) (to appear). (In Russian)

Steklov Math.Inst.Ac.Sc.
Moscow GSP-1, 117966
Vavilova 42

ON WAVELET ANALYSIS

Charles K. Chui

The objective of this paper is to introduce and classify various classes
of wavelets and to discuss some of their basic properties. In particular,
an identity of Littlewood-Paley type is given and is used to partially
characterize wavelets and their duals.

1. Introduction

There are two main components in Fourier Analysis, namely, Fourier
series and integral Fourier transform. These two types of "transforms" are
unrelated since they are defined on two different classes of functions: periodic
functions and functions defined on the real line $\mathbb{R}$, respectively. Recently, the
subject of "wavelets" has become a very attractive area of research among
both pure and applied mathematicians. Depending on the expertise or interests of an individual researcher, it may be considered as a topic in harmonic
analysis, operator theory, applied mathematics, etc. The objective of this paper is to express author's point of view, that this area of research is rich and
important enough to be called "Wavelet Analysis", a subject title which has
already been used by several experts in this area. Indeed, similar to Fourier
Analysis, there are also two main components in Wavelet Analysis, namely:
wavelet series and integer wavelet transform. However, in contrast to Fourier
Analysis, these two "transforms" are intimately related, both being defined
on the same function classes whose domain of definition is $\mathbb{R}$. In fact, the
coefficient sequence of a wavelet series with basic wavelet ψ is the sequence
of values of the integral wavelet transform with respect to a wavelet $\tilde{\psi}$, which
is dual to ψ. These values are evaluated at the binary scaled levels and
their corresponding dyadic points. Of course, there are other variations of
wavelets, such as those defined on bounded domains, locally compact abelian
groups, etc.; but we will only consider functions in $L^2 := L^2(\mathbb{R})$, although
extensions to $L^2(\mathbb{R}^s)$, $s > 1$, are very interesting, particularly to the approximation theorists, since multivariate wavelets can be constructed by using

Supported by NSF Grant #DMS-89-0-01345 and ARO Contract #DAAL 03-90-G-0091.

box-splines (cf. [5,8,21,23]).

In order to be more focused, no attempt is being made to give a comprehensive introduction of this subject, even from this somewhat narrow point of view. The interested reader is referred to the forthcoming book [2] by this author. More advanced monographs with different emphases are or will be available (cf. Meyer [27] and Daubechies [17]). There are also volumes of edited papers, such as [1,3,14,22]; and a new book series, entitled "Wavelet Analysis and Its Applications" is being launched by Academic Press, Inc. at Boston.

The outline of this paper is as follows. Section 2 will be devoted to a discussion of frames and Riesz bases. The concept of integral wavelet transforms (IWT) and methods for finding their inverses are studied in Section 3. Wavelets and their classification are discussed in Section 4; and in Section 5, Littlewood-Paley type identities for frames and wavelets are introduced. Such identities are also used to partially characterize wavelets, particularly the so-called "dyadic wavelets".

2. Frames and Riesz bases

A family of functions $\phi_k \in L^2$ is said to be a frame of L^2 if there exist two positive constants c_1 and c_2 with $0 < c_1 \leq c_2 < \infty$, such that

$$c_1\|f\|^2 \leq \sum_{k \in \mathbb{Z}} |\langle f, \phi_k \rangle|^2 \leq c_2\|f\|^2,$$

for all $f \in L^2$. Here, $\langle \cdot, \cdot \rangle$ denotes the inner product and $\| \cdot \|$ the norm of L^2. The notion of frames was introduced by Duffin and Schaeffer [18] in their work on nonharmonic analysis, and a comprehensive study was given by Daubechies [16,17]. Although this notion is quite general, we will only restrict our attention to the family of functions $\psi_{j,k}$ generated by a single function $\psi \in L^2$ as follows:

$$\psi_{j,k}(x) := 2^{\frac{j}{2}} \psi(2^j x - k), \qquad j, k \in \mathbb{Z}. \tag{2.1}$$

Hence, $\mathcal{F} = \{\psi_{j,k} : j, k \in \mathbb{Z}\}$ is a frame of L^2 if positive constants A and B exist, with $0 < A \leq B < \infty$, such that

$$A\|f\|^2 \leq \sum_{j,k} |\langle f, \psi_{j,k} \rangle|^2 \leq B\|f\|^2, \tag{2.2}$$

for all $f \in L^2$. The constants A and B are called frame bounds of the family $\mathcal{F}$. Recall that the same family $\mathcal{F}$ is called a Riesz basis of L^2 if there exist constants C and D, with $0 < C \leq D < \infty$, such that

$$C \sum_{j,k} |c_{j,k}|^2 \leq \left\| \sum_{j,k} c_{j,k} \psi_{j,k} \right\|^2 \leq D \sum_{j,k} |c_{j,k}|^2, \tag{2.3}$$

for all sequences $\{c_{j,k}\} \in \ell^2 := \ell^2(\mathbf{Z}^2)$. We will call the constants C and D Riesz bounds of the family. The following observation is clear.

PROPOSITION 2.1. *Any Riesz basis $\{\psi_{j,k}\}$ of L^2 generated as in (2.1) by an L^2 function ψ is also a frame of L^2. Furthermore, the frame bounds are also given by the Riesz bounds.*

A simple derivation of the above statement was given in [6] as follows. Let $\{\psi^{j,k}\}$, $j, k \in \mathbf{Z}$, denote the dual of $\{\psi_{j,k}\}$, in the sense of

$$\langle \psi_{j,k}, \psi^{j',k'} \rangle = \delta_{j,j'} \delta_{k,k'}. \tag{2.4}$$

Then $\{\psi^{j,k}\}$ is clearly a Riesz basis of L^2 with Riesz bounds D^{-1} and C^{-1}. By the duality (2.4), we may write

$$f(x) = \sum_{j,k} \langle f, \psi_{j,k} \rangle \psi^{j,k}(x).$$

Hence, from the notion of Riesz basis and Riesz bounds, we have

$$D^{-1} \sum_{j,k} |\langle f, \psi_{j,k} \rangle|^2 \le \|f\|^2 \le C^{-1} \sum_{j,k} |\langle f, \psi_{j,k} \rangle|^2$$

by considering the ℓ^2 sequence $c_{j,k} := \langle f, \psi_{j,k} \rangle$. This is the same as (2.2) with $A = C$ and $B = D$, completing the proof of the Proposition.

It is important to point out that the dual basis $\{\psi^{j,k}\}$, relative to the Riesz basis $\{\psi_{j,k}\}$, may not be formulated as binary dilations and integral translations of a single function, as the formulation of $\{\psi_{j,k}\}$ in (2.1). A class of counterexamples is given in Section 4, where the definition and a classification of wavelets are introduced.

On the other hand, the class of frames described by (2.2) is larger than the class of Riesz bases described by (2.3). A typical example, constructed by Daubechies [17] is the Haar-like function $\psi_h = \chi_{[0,3)} - \chi_{[3,6)}$. Daubechies [16,17] has also shown that even every frame $\{\psi_{j,k}\}$ as in (2.1) has a dual frame $\{\psi^{j,k}\}$ defined by

$$\langle f, g \rangle = \sum_{j,k} \langle f, \psi_{j,k} \rangle \langle \overline{\psi^{j,k}, g} \rangle, \qquad f, g \in L^2,$$

but already observed that $\{\psi^{j,k}\}$ may not come from one single function.

For any frame considered in this paper, the following result was established in [6].

2 Тр. конференции

THEOREM 2.1. *Let $\psi \in L^2$ generate a frame $\{\psi_{j,k}\}$ of L^2 with frame bounds A and B as defined in (2.1) and (2.2). Then the Fourier transform $\widehat{\psi}$ of ψ satisfies*

$$A \leq \sum_{j \in \mathbb{Z}} |\widehat{\psi}(2^j \omega)|^2 \leq B, \quad \text{a.e., and} \tag{2.5}$$

consequently,

$$2A \log 2 \leq \int_{-\infty}^{\infty} \frac{|\widehat{\psi}(\omega)|^2}{|\omega|} d\omega \leq 2B \log 2. \tag{2.6}$$

To derive (2.6) from (2.5), we simply integrate each quantity in the following inequalities:

$$\frac{A}{|\omega|} \leq \sum_{j \in \mathbb{Z}} \frac{|\widehat{\psi}(2^j \omega)|^2}{|\omega|} \leq \frac{B}{|\omega|}$$

over the set $\{\omega: 1 \leq |\omega| \leq 2\}$, and notice that

$$\sum_{j \in \mathbb{Z}} \int_1^2 \frac{|\widehat{\psi}(2^j \omega)|^2}{|\omega|} d\omega = \sum_{j \in \mathbb{Z}} \int_{2^j}^{2^{j+1}} \frac{|\widehat{\psi}(\omega)|^2}{|\omega|} d\omega = \int_0^{\infty} \frac{|\widehat{\psi}(\omega)|^2}{|\omega|} d\omega,$$

and that

$$\sum_{j \in \mathbb{Z}} \int_{-2}^1 \frac{|\widehat{\psi}(2^j)|^2}{|\omega|} d\omega = \int_{-\infty}^0 \frac{|\widehat{\psi}(\omega)|^2}{|\omega|} d\omega.$$

This not only yields (2.6), but also enables us to conclude that each of the integrals over $(-\infty, 0)$ and $(0, \infty)$ lies between $A \log 2$ and $B \log 2$.

We remark that (2.6) was already derived by Daubechies [16] by using a different method. It is also interesting to observe that if $\widehat{\psi}$ is continuous at 0, then $\widehat{\psi}(0) = 0$, or equivalently

$$\int_{-\infty}^{\infty} \psi(x) dx = 0. \tag{2.7}$$

Hence, the graph of ψ looks like a small wave. The continuity of $\widehat{\psi}$ is, of course, a consequence of the assumption $\psi \in L^1 \cap L^2$. In view of Proposition 2.1, any ψ that generates a Riesz basis of L^2 satisfies (2.5) and (2.6).

3. Integral wavelet transforms and their inverses

We now make a digression to consider yet another topic, called "integral wavelet transform" (or IWT). To do so, we need a kernel function $\psi \in L^2$, which is usually called a "mother wavelet". Then the IWT of an $f \in L^2$, as introduced by Grossmann and Morlet [20], is defined to be

$$(W_\psi f)(x, y) = \frac{1}{\sqrt{y}} \int_{-\infty}^{\infty} f(t)\overline{\psi\left(\frac{t - x}{y}\right)} dt, \qquad (3.1)$$

where $y > 0$ and $x \in \mathbb{R}$. Why is this transform, which is really very simple, so special? In applications, if ψ has very fast decay, or even better, compact support, then $W_\psi f$ "windows" the function f in a neighborhood of $x + x^*$, where x^* is the "center" of the kernel ψ. This is called "time-localization", when f is considered as an "analog signal". If ψ is so chosen that its Fourier transform $\hat{\psi}$ also has very fast decay, then by setting $\eta(\omega) := \overline{\hat{\psi}(\omega + \omega^*)}$, where ω^* is the "center" of $\hat{\psi}$, and by observing that

$$(W_\psi f)(x, y) \equiv \frac{\sqrt{y}}{2\pi} \int_{-\infty}^{\infty} \hat{f}(\omega)e^{ix\omega}\eta\left(y\left(\omega - \frac{\omega^*}{y}\right)\right) d\omega, \qquad (3.2)$$

we see that the localized-time information on the left-hand side of (3.2) is identical to the localized-frequency information on the right-hand side of (3.2). The importance of the "Window inverse Fourier transform" in (3.2) is that the sizes of both the time-window on the left and the frequency-window on the right are flexible although their product, which is the area of the time-frequency window, is a constant. By calibrating the frequency-axis as a constant multiple of the reciprocal of the scale parameter y, we have a constant-Q bandpass filter, meaning that the quotient of the center frequency with the width of frequency-window is unchanged with any shift of the center frequency. This is the so-called zoom-in and zoom-out effect of the IWT. For more details, the interested reader is referred to [20,15,2,4,17].

We next discuss the inverse IWT's; that is, we are interested in reconstructing f from values of $(W_\psi f)(x, y)$. Let us classify this problem into three types, as follows.

3.1. Recovery from $(W_\psi f)(x, y)$ where $y > 0$ and $x \in \mathbb{R}$

This was done by Grossmann and Morlet [20], where they must assume that

$$c_\psi := \int_{-\infty}^{\infty} \frac{|\hat{\psi}(\omega)|^2}{|\omega|} d\omega < \infty. \qquad (3.3)$$

(Cf. (2.6)-(2.7) and the comment in Section 2 on ψ being a "small wave".) The reconstruction formula in [20] is given by

$$f(t) = \frac{1}{c_\psi} \iint_{\mathbb{R}^2} (W_\psi f)(x, y)\frac{1}{|y|^{1/2}}\psi\left(\frac{x - t}{y}\right) \frac{dxdy}{y^2}, \qquad (3.4)$$

2*

where we could think of ψ as a "dual" to itself. Of course, this dual cannot be unique. For time-frequency analysis where only positive ω is allowed, the integral is (3.4) may be taken over $\mathbb{R} \times (0, \infty)$, provided both ψ and f are real-valued and the integral in the definition of c_ψ in (3.3) is also taken on $(0, \infty)$, (see Daubechies [16]).

3.2. *Recovery from* $(W_\psi f)(x, 2^{-j})$, $j \in \mathbb{Z}$ *and* $x \in \mathbb{R}$

This is very useful in applications to data compression, especially for images (cf. [25,26]), where the "mother wavelet" ψ is called a *dyadic wavelet*. The reconstruction formula given in [25] is

$$f(t) = \sum_{j \in \mathbb{Z}} \int_{-\infty}^{\infty} (W_\psi f)(x, 2^{-j}) 2^j \psi^*(2^j(t - x)) dx, \qquad (3.5)$$

where ψ^* is a "dual" of ψ. In Section 5, we will discuss a characterization of ψ^* in terms of a so-called Littlewood-Paley identity, and observe, in particular, the non-uniqueness of ψ^*, in general. If ψ satisfies the "stability" condition (2.5) (so that (2.6) is again satisfied), then a very convenient dual used by Mallet and Hwang [25] is χ whose Fourier transform is given by

$$\hat{\chi}(\omega) = \frac{\overline{\hat{\psi}(\omega)}}{\sum_{j \in \mathbb{Z}} |\hat{\psi}(2^j \omega)|^2}. \qquad (3.6)$$

For more details, the reader is referred to [25,26].

3.3. *Recovery from* $(W_\psi f)(\frac{k}{2^j}, 2^{-j})$, $j, k \in \mathbb{Z}$

This is probably the most interesting, especially from the mathematical point of view. In addition, very efficient algorithms are available (cf. [9,10,11,15,24]). The reconstruction formula is

$$f(t) = \sum_{j,k \in \mathbb{Z}} \left\{ (W_\psi f) \left(\frac{k}{2^j}, 2^{-j} \right) \right\} 2^{j/2} \tilde{\psi}(2^j t - k), \qquad (3.7)$$

where $\tilde{\psi}$ is the "dual" of ψ. This time the mathematical problem is existence but *not* uniqueness; since if $\{\psi_{j,k}\}$ as defined in (2.1) is a Riesz basis of L^2, then it dual $\{\psi^{j,k}\}$ is always unique. But as we will see in the next section, there may *not* exist a $\tilde{\psi} \in \mathbb{R}^2$ such that

$$\psi^{j,k}(t) = 2^{j/2} \tilde{\psi}(2^j t - k). \qquad (3.8)$$

4. Wavelets and classification of wavelets

Let $\psi \in L^2$ and $\psi_{j,k}$ be defined by (2.1). In this section, we will always assume that $\{\psi_{j,k}\}$, $j, k \in \mathbb{Z}$, is a Riesz basis of L^2. So, $\{\psi_{j,k}\}$ has a unique dual $\{\psi^{j,k}\}$ defined by (2.4). We will call ψ a *wavelet* (or more precisely, a "mother wavelet" for the IWT: $(W_\psi f)\left(\frac{k}{2^j}, 2^{-j}\right)$), if there exists a $\tilde{\psi} \in L^2$ that determines the dual basis $\{\psi^{j,k}\}$ as in (3.8). The most commonly known wavelets are "orthonormal wavelets" (or simply, o.n. wavelets), which are of the most restrictive type in our following classification.

4.1. *Orthonormal wavelets (or o.n. wavelets)*

ψ is called an *o.n. wavelet* if the Riesz basis $\{\psi_{j,k}\}$ is an o.n. basis of L^2. Since there is a lot written on this topic, we are not going into it but only refer the interested reader to the fundamental paper of Daubechies [17], as well as the monographs [2,17,27]. Note that since $\{\psi_{j,k}\}$ is o.n., it is self-dual; that is, $\psi^{j,k} = \psi_{j,k}$, so that $\tilde{\psi} \equiv \psi$.

If the Riesz basis $\{\psi_{j,k}\}$ is not o.n., however, then finding the dual basis $\{\psi^{j,k}\}$ is usually not simple, and in fact, $\{\psi^{j,k}\}$ may no longer come from a single function as stated in the following.

PROPOSITION 4.1. *Let $\{\psi_{j,k}\}$, as defined in (2.1), be an o.n. basis of L^2, and consider*

$$\begin{cases} \eta_z(x) := \psi(x) - \bar{z}\sqrt{2}\psi(2x), & |z| < 1; \\ \eta_{z;j,k}(x) := 2^{\frac{j}{2}}\eta_z(2^j x - k). \end{cases} \tag{4.1}$$

Then $\{\eta_{z;j,k}\}$, $j, k \in \mathbb{Z}$, is a Riesz basis of L^2, but its dual $\{\eta^{j,k}_z\}$ is not generated by some $\tilde{\eta}_z \in L^2$ as in (3.8), at least for almost all z in $|z| < 1$.

In the following, we present a simple proof given in [6]. That $\{\eta_{z;j,k}\}$, $j, k \in \mathbb{Z}$, is a Riesz basis of L^2 is clear; and in fact, we have

$$(1 - |z|^2)^2 \sum_{j,k} |a_{j,k}|^2 \leq \left\| \sum_{j,k} a_{j,k}\eta_{z;j,k} \right\|^2$$
$$\leq (1 + |z|^2)^2 \sum_{j,k} |a_{j,k}|^2,$$

for all $\{a_{j,k}\} \in \ell^2(\mathbb{Z}^2)$, by using the orthonormality condition of $\{\psi_{j,k}\}$. In addition, by direct computation, it is easy to see that·

$$\begin{cases} \eta^{0,1}_z(z) = \psi_{0,1}(x) \quad \text{and} \\ \eta^{0,0}_z(z) = \sum_{\ell=0}^{\infty} \psi_{-\ell,0}(x)z^\ell. \end{cases}$$

Hence, if there exists some $\bar{\eta}_z \in L^2$ such that

$$\eta_z^{j,k}(x) = 2^{\frac{j}{2}}\bar{\eta}_z(2^j x - k) =: \bar{\eta}_{z;j,k}(x). \tag{4.2}$$

Then we have:

$$\psi(x) = \eta_z^{0,1}(x+1) = \bar{\eta}_{z;0,1}(x+1) = \bar{\eta}_{z;0,0}(x) = \eta_z^{0,0}(x) = \sum_{\ell=0}^{\infty} \psi_{-\ell,0}(x)z^{\ell},$$

so that

$$\sum_{\ell=1}^{\infty} \psi_{-\ell,0}(x)z^{\ell} = 0, \qquad 0 < |z| < 1.$$

That this is absurd completes the proof of the Proposition.

Recall that ψ is called a wavelet if it has a *dual* $\tilde{\psi}$, in the sense of $\psi^{j,k} = \tilde{\psi}_{j,k}$. Hence, the dual $\tilde{\psi}$ of a wavelet ψ is also a wavelet, with the dual given by ψ. That is, the wavelet series (3.7) can be written as

$$f = \sum_{j,k}\langle f, \psi_{j,k}\rangle \tilde{\psi}_{j,k} = \sum_{j,k}\langle f, \tilde{\psi}_{j,k}\rangle \psi_{j,k}, \tag{4.3}$$

where the coefficients in (4.3) are the IWT's of f evaluated at $(k \cdot 2^{-j}, 2^{-j})$ by using ψ and $\tilde{\psi}$, respectively, as the mother wavelets. Proposition 4.1 says that not all Riesz bases are generated by wavelets. For many applications, the o.n. property has to be sacrificed in order to achieve certain desirable specifications. This leads to the second type of wavelets, to be discussed next.

4.2. *Semi-orthogonal wavelets (or s.o. wavelets)*

ψ is called an *s.o. wavelet* if the Riesz basis $\{\psi_{j,k}\}$ satisfies:

$$\langle \psi_{j,k}, \psi_{\ell,m}\rangle = 0, \quad j \neq \ell \text{ and } j,k,\ell,m \in \mathbb{Z}. \tag{4.4}$$

To see why an s.o. wavelet is a wavelet, we simply verify that $\tilde{\psi}$ defined by

$$\hat{\tilde{\psi}}(\omega) := \frac{\hat{\psi}(\omega)}{\sum_{k} |\hat{\psi}(\omega + 2\pi k)|^2} \tag{4.5}$$

is the dual of ψ.

Probably the most useful s.o. wavelets are the spline-wavelets introduced in [10], which we will describe next.

Let m be any positive integer and denote by N_m the mth order B-spline with integer knots, unit integral, and support given by $[0, m]$; that is, N_m is the m-fold convolution of the characteristic function of the unit interval $[0,1)$. Then the mth order B-spline-wavelet (or simply, m^{th} order B-wavelet) constructed in [10] has the explicit formula:

$$\psi_m(x) = \frac{1}{2^{m-1}} \sum_{j=0}^{2m-2} (-1)^j N_{2m}(j+1) N_{2m}^{(m)}(2x - j) \qquad (4.6)$$

$$= \sum_{j=0}^{3m-2} q_j^{(m)} N_m(2x - j),$$

where

$$q_j^{(m)} := \frac{(-1)^j}{2^{m-1}} \sum_{\ell=0}^{m} \binom{m}{\ell} N_{2m}(j - \ell + 1). \qquad (4.7)$$

To describe some of the important properties of ψ_m, let us consider the spaces

$$W_j := \text{clos}_{L^2} \langle \psi_{m;j,k} : k \in \mathbf{Z} \rangle,$$

where, as usual, the notation of the L^2-closure of a linear span is used, and $\psi_{m;j,k}(x) := 2^{-j/2} \psi_m(2^j x - k)$. The proof of the following result can be found in [10,11].

THEOREM 4.1. *Let m be an arbitrarily given positive integer. Then ψ_m is an s.o. wavelet in the sense that $\{\psi_{m;j,k} : j, k \in \mathbf{Z}\}$ is a Riesz basis of L^2 and $W_j \perp W_\ell$ for all $j \neq \ell$, $j, \ell \in \mathbf{Z}$. In other words,*

$$L^2 = \bigoplus_{j \in \mathbf{Z}} W_j,$$

where $\oplus$ denotes the orthogonal sum. Furthermore, ψ_m has minimal support in W_0, and the dual of ψ_m is given by

$$\widetilde{\psi}_m(x) = \frac{(-1)^{m+1}}{2^{m-1}} \sum_j \alpha_j^{(m)} L_{2m}^{(m)}(2x - 2m + 1 - 2j), \qquad (4.8)$$

where L_{2m} is the $(2m)^{\text{th}}$ order spline function with integer knots determined by the fundamental property: $L_{2m}(j) = \delta_{j,0}$, $j \in \mathbf{Z}$; and the coefficient sequence $\{\alpha_j^{(m)}\}$ in (4.8) is the determining coefficient sequence in the B-spline series expansion of L_{2m}, namely:

$$L_{2m}(x) = \sum_j \alpha_j^{(m)} N_{2m}(x + m - j).$$

Finally, the third type of wavelets is the collection of those which are not s.o. wavelets. We classify them as *non-orthogonal wavelets*. However, being a wavelet, ψ has a dual $\tilde{\psi}$, and by this we mean:

$$\langle \psi_{j,k}, \tilde{\psi}_{\ell,m} \rangle = \delta_{j,\ell} \delta_{k,m};$$

and this is why a non-orthogonal wavelet is also called a bi-orthogonal wavelet in the literature [12,13,19].

5. Littlewood-Paley identities

If ψ is an o.n. wavelet, then its Fourier transform satisfies the identity:

$$\sum_{j \in \mathbb{Z}} |\hat{\psi}(2^j \omega)|^2 = 1, \quad \text{a.e.,} \tag{5.1}$$

called the Littlewood-Paley identity (cf. [17,27]). In [7], this result is generalized to any wavelet, as follows.

THEOREM 5.1. Let $\psi \in L^2$ be a wavelet with dual $\tilde{\psi}$. Then

$$\operatorname*{ess\,sup}_{\omega \in \mathbb{R}} \sum_{j \in \mathbb{Z}} |\hat{\psi}(2^j \omega) \hat{\tilde{\psi}}(2^j \omega)| < \infty, \tag{5.2}$$

and

$$\sum_{j \in \mathbb{Z}} \overline{\hat{\psi}(2^j \omega)} \hat{\tilde{\psi}}(2^j \omega) = 1, \quad \text{a.e.} \tag{5.3}$$

Recall from Section 2 the notion of dual frames. If the dual frame $\{\psi^{j,k}\}$ of a frame $\{\psi_{j,k}\}$ is also generated by some $\tilde{\psi} \in L^2$ as in (3.8), then we call ψ a wavelet-frame and $\tilde{\psi}$ its corresponding dual wavelet-frame. We remark that the proof of Theorem 5.1 in [7] also applies to wavelet-frames. Also recall from Section 3 the notion of dyadic wavelets ψ with dual ψ^*, as described by (3.5). In [7], the following result is also established.

THEOREM 5.2. Let ψ be a dyadic wavelet satisfying the stability condition (2.5). Suppose that ψ^* is a function in L^2 satisfying

$$\operatorname*{ess\,sup}_{\omega \in \mathbb{R}} \sum_{j \in \mathbb{Z}} |\hat{\psi}^*(2^j \omega)|^2 < \infty.$$

Then ψ^* is a dual of ψ in the sense of (3.5) if and only if the Littlewood-Paley identity:

$$\sum_{j \in \mathbb{Z}} \overline{\hat{\psi}(2^j \omega)} \hat{\psi}^*(2^j \omega) = 1, \quad \text{a.e.,}$$

is satisfied.

However, in general, the dual ψ^* of a dyadic wavelet ψ is not unique. In fact, if a dyadic wavelet ψ is also an s.o. wavelet, then it follows from Theorem 5.2 that both χ_1 and χ_2, defined by

$$\widehat{\chi}_1(\omega) := \frac{\overline{\widehat{\psi}(\omega)}}{\sum_{j \in \mathbb{Z}} |\widehat{\psi}(2^j \omega)|^2} \tag{5.4}$$

and

$$\widehat{\chi}_2(\omega) := \frac{\overline{\widehat{\psi}(\omega)}}{\sum_{k \in \mathbb{Z}} |\widehat{\psi}(\omega + 2\pi k)|^2}, \tag{5.5}$$

are duals of the dyadic wavelet ψ. Of course, if ψ is an o.n. wavelet, then both of the denominators in (5.4) and (5.5) are equal to 1, so that $\chi_1 = \chi_2$ a.e.. But in general, when ψ is s.o. but not o.n., then $\chi_1 \neq \chi_2$. This is certainly the case for the B-wavelet ψ_m in (4.6) for any $m \geq 2$.

References

[1] G. Beylkin, R. Coifman, I. Daubechies, S. Mallat, Y. Meyer, L. Raphael, and B. Ruskai, eds., *Wavelets and Their Applications*, Jones and Bartlett, Boston (to appear).

[2] C.K. Chui, *Introduction to Wavelets*, Academic Press, Boston, 1992.

[3] C.K. Chui, ed., *Wavelets: a Tutorial*, Academic Press, Boston, 1992.

[4] C.K. Chui, *Wavelets: with Emphasis on Time-Frequency Analysis*, Under preparation..

[5] C.K. Chui and C. Li, *A general framework of multivariate compactly supported wavelets and dual wavelets*, Under preparation..

[6] C.K. Chui and X.L. Shi, *Inequalities of Littlewood-Paley type for frames and wavelets*, CAT Report #249, Texas A&M University, 1991.

[7] C.K. Chui and X.L. Shi, *On a Littlewood-Paley identity and characterization of wavelets*, CAT Report #250, Texas A&M University, 1991.

[8] C.K. Chui, J. Stöckler, and J.D. Ward, *Compactly supported boxspline wavelets*, CAT Report #230, Texas A&M University, 1990.

[9] C.K. Chui and J.Z. Wang, *A cardinal spline approach to wavelets*, Proc. Amer. Math. Soc, (to appear in 1991).

[10] C.K. Chui and J.Z. Wang, *On compactly supported spline wavelets and a duality principle*, Trans. Amer. Math. Soc, (to appear in 1991).

[11] C.K. Chui and J.Z. Wang, *A general framework of compactly supported splines and wavelets*, CAT Report #210, Texas A&M University, 1991.

[12] A. Cohen, Doctoral Thesis, Univ. Paris-Dauphine, 1990.

[13] A. Cohen, I. Daubechies, and J.C. Feauveau, *Biorthogonal bases of compactly supported wavelets*, Comm. Pure and Appl. Math. (to appear).

[14] J.M. Combes, A. Grossmann, and Ph. Tchamitchian, eds., *Wavelets: Time-Frequency Methods and Phase Space*, Springer-Verlag, N.Y., 1989; 2nd Edition, 1991.

[15] I. Daubechies, *Orthonormal bases of compactly supported wavelets*, Comm. Pure and Appl. Math. **41** (1988), pp. 909–996.

[16] I. Daubechies, *The wavelet transform, time-frequency localization and signal analysis*, IEEE Trans. Information Theory **36** (1990), pp. 961–1005.

[17] I. Daubechies, *Ten Lectures on Wavelets*, CBMS-NSF Series in Applied Math., SIAM Publ., Philadelphia (to appear).

[18] R.J. Duffin and A.C. Schaeffer, *A class of nonharmonic Fourier series*, Trans. Amer. Math. Soc. **72** (1952), pp. 341–366.

[19] J.C. Feauveau, *Non-orthogonal multiresolution analysis using wavelets*, Wavelets: a Tutorial, ed. by C.K. Chui,, Academic Press, Boston, 1992.

[20] A. Grossmann and J. Morlet, *Decomposition of Hardy functions into integrable wavelets of constant shape*, SIAM J. Math. Analysis **15** (1984), pp. 723–736.

[21] R.Q. Jia and C.A. Micchelli, *Using the refinement equation for the construction of pre-wavelets V: extensibility of trigonometric polynomials*, Preprint, 1991.

[22] W. Light, ed., *Wavelets, Subdivisions, and Radial Functions*, Oxford University Press, Oxford, 1991.

[23] R. Lorentz and W. Madych, *Wavelets and generalized box splines*, Preprint, 1991.

[24] S. Mallat, *Multiresolution approximations and wavelets orthonormal bases of $L^2(\mathbb{R})$*, Trans. Amer. Math. Soc. **315** (1989), pp. 69–87.

[25] S. Mallat and W.L. Hwang, *Singularity detection and processing with wavelets*, Preprint, 1991.

[26] S. Mallat and S. Zhong, *Wavelet transform maxima and multiscale edges*, Wavelets and Their Applications (G. Beylkin et. al., Jones and Bartlett, eds.), Boston (to appear).

[27] Y. Meyer, *Ondelettes et Opérateurs*, (two volumes), Hermann, Paris, 1990.

Center for Approximation Theory
Department of Mathematics
Texas A&M University
College Station
TX 77843, U.S.A.

POLYNOMIALS ORTHOGONAL ON THE UNIT CIRCLE WITH RANDOM RECURRENCE COEFFICIENTS

J.S. Geronimo

Polynomials orthogonal on the unit circle whose recurrence coefficients are generated from a stationary stochastic process are considered. A Lyapunov exponent introduced and its properties are related to absolutely continuous components of the orthogonality measure.

1. Introduction

Let $\{\phi(z,n)\}_{n=0}^{\infty}$ be a sequence of polynomials orthonormal with respect to measure σ supported on the unit circle. It is well known (Szegö [17]) that these polynomials satisfy the following recurrence formula

$$\phi(z,n) = a(n)(z\phi(z,n-1) + \alpha(n)\phi^*(z,n-1)) \tag{1.1}$$

and

$$\phi^*(z,n) = a(n)(\phi^*(z,n-1) + \overline{\alpha}(n)z\phi(z,n-1)) \tag{1.2}$$

Here we will consider the above equations when $\{\alpha(n)\}_{n=1}^{\infty}$ is a stationary stochastic process. Later we (Geronimo, Johnson [6]) will take up the problem when the process is extended to negative infinity. The results given below were inspired by similar results proved by Kotani [9] who considered stochastic Schrödinger equations and by Simon [14] and Craig and Simon [2] who considered discrete analogs of the Schrödinger equations. For other results related to random polynomials orthogonal on the unit circle see Nikishin [11] and Teplyaev [16].

We proceed as follows. In Section 2 we introduce the necessary notation and solutions to the difference equations. In Section 3 we consider the recurrence coefficients as a stationary stochastic process and prove the existence and nonnegativity of the Lyapunov exponent. Also in this section we relate the Lyapunov exponent to a certain non–random measure. This is the analog

Supported in part by NSF grant DMS-9005944

of the so–called Thouless formula [2]. In Section 4 m functions are introduced
and used to relate the nonexistence of the absolutely continuous components
of the orthogonality measure to the positivity of radial boundary values of
the Lyapunov exponent. Finally in Section 5 we consider the case when the
coefficients are periodic.

2. Preliminaries

We begin by writing (1.1) and (1.2)

$$X(z,n) = T_n(z)X(z,n-1), \tag{2.1}$$

where

$$T_n(z) = a(n) \begin{pmatrix} z & \alpha(n) \\ \overline{\alpha(n)}z & 1 \end{pmatrix}, \tag{2.2}$$

with $\alpha(n) \in \mathbb{C}$ and $|\alpha(n)| < 1$, for $n = 1, 2, \ldots$, and $a(n) > 0$ $n = 0, 1, 2, \ldots$.
$a(n)$ is related to $\alpha(n)$ by the equation

$$\frac{1}{a(n)^2} = 1 - |\alpha(n)|^2. \tag{2.3}$$

From (2.3) it follows that $\det T_n(z) = z$. If we set

$$T_{i,k}(z) = T_k(z)T_{k-1}(z)\ldots T_i(z), \tag{2.4}$$

then $\det T_{1,k} = z^k$, which implies that $T_{1,k}$ is a fundamental matrix solution
to the difference equation (2.1). If we choose the vector $\Phi(z,0) = \binom{1}{1}$ then

$$\Phi(z,n) = \begin{pmatrix} \phi(z,n) \\ \phi^*(z,n) \end{pmatrix} = T_{1,n}(z) \begin{pmatrix} 1 \\ 1 \end{pmatrix}, \tag{2.5}$$

where $\phi^*(z,n) = z^n\overline{\phi}(\frac{1}{z},n)$ gives the well known Szegö polynomials i.e.,
$\phi(z,n)$ is a polynomial of degree n in z with leading coefficient positive and
such that

$$\int_{-\pi}^{\pi} \phi(e^{i\theta},n)e^{-ik\theta}d\sigma = 0, \quad 0 \le k < n, \tag{2.6}$$

for some positive measure σ.

Another useful solution is

$$\Phi_1(z,n) = \begin{pmatrix} \phi_1(z,n) \\ \phi_2(z,n) \end{pmatrix} = T_{1,n}(z) \begin{pmatrix} 1 \\ -1 \end{pmatrix}, \tag{2.7}$$

and it is not difficult to show by induction that

$$\phi_2(z,n) = -\phi_1(z,n)^*. \tag{2.8}$$

If we write

$$T_{1,n}(z) = \begin{pmatrix} \psi_1(z,n) & \Gamma_1(z,n) \\ \psi_2(z,n) & \Gamma_2(z,n) \end{pmatrix}, \tag{2.9}$$

then

$$\Psi(z,n) = \begin{pmatrix} \psi_1(z,n) \\ \psi_2(z,n) \end{pmatrix} = T_{1,n} \begin{pmatrix} 1 \\ 0 \end{pmatrix}, \tag{2.10}$$

and

$$\Gamma(z,n) = \begin{pmatrix} \Gamma_1(z,n) \\ \Gamma_2(z,n) \end{pmatrix} = T_{1,n} \begin{pmatrix} 0 \\ 1 \end{pmatrix}, \tag{2.11}$$

and it follows that

$$\Psi(z,n) = \frac{1}{2}\left[\Phi(z,n) + \Phi_1(z,n)\right], \tag{2.12}$$

and

$$\Gamma(z,n) = \frac{1}{2}\left[\Phi(z,n) - \Phi_1(z,n)\right]. \tag{2.13}$$

The Wronskian W of any two solutions of (2.1) is defined to be

$$W[X_1(z,n), X_2(z,n)] = \det[X_1(z,n), X_2(\cdot,n)], \tag{2.14}$$

and using (2.1) we find that $W[X_1(z,n), X_2(z,n)] = zW[X_1(z,n-1), X_2(z, n-1)]$. This implies that $\Phi(z,n)$ and $\Phi_1(z,n)$ are linearly independent solutions of (2.1) except at z equal to zero.

LEMMA 2.1. Let $|\alpha(i)| < 1$ for $i = 1, 2, \ldots, n$, and set

$$z^{\frac{n}{2}}w_1(z) = \frac{(\phi(z,n) + \phi^*(z,n))}{2}, \quad z^{\frac{n}{2}}w_2(z) = \frac{(\phi_1(z,n) - \phi_2(z,n))}{2},$$

$$z^{\frac{n}{2}}w_3(z) = \frac{(\phi(z,n) - \phi^*(z,n))}{2}, \quad z^{\frac{n}{2}}w_4(z) = \frac{(\phi_1(z,n) + \phi_2(z,n))}{2}.$$

Then $\phi^*(z,i)$ and $\phi_2(z,i)$, $i = 1, 2, \ldots, n$, have all their zeros located outside the closed unit disk. Furthermore for z on the unit circle w_1 and w_2 are real while w_3 and w_4 are purely imaginary. The zeros of $w_i, i = 1, \ldots, 4$ lie on the unit circle with the zeros of $z^{\frac{n}{2}}w_1(z)$ interlacing those of $z^{\frac{n}{2}}w_3(z)$ and the zeros of $z^{\frac{n}{2}}w_2(z)$ interlacing those of $z^{\frac{n}{2}}w_4(z)$. Finally $w_1 w_2 - w_3 w_4 = 1$

which implies that the zeros of $z^{\frac{n}{2}} w_1(z)$ interlace those of $z^{\frac{n}{2}} w_4(z)$ while the zeros of $z^{\frac{n}{2}} w_2(z)$ interlace those of $z^{\frac{n}{2}} w_3(z)$.

Proof. It is a well known fact that if $|\alpha(i)| < 1$ for $i = 1, 2, ..., n$, then $\phi^*(z, i)$, $i = 1, 2, ..., n$ has all its zeros outside the closed unit disk (Geronimus [7]). From the recurrence relation for $\phi_2(z, i)$, (2.7) we find that

$$\phi_2(z, i) = a(i)(\phi_2(z, i - 1) + \overline{\alpha(i)} z \phi_1(z, i - 1)). \tag{2.15}$$

If $|\phi_2(z, i-1)| > 0$ for $|z| \leq 1$, then (2.8) shows that $|\frac{\phi_1(z, i-1)}{\phi_2(z, i-1)}| \leq 1$ for $|z| \leq 1$ which implies via the above recurrence formula that the zeros of $\phi_2(z, i)$ are located outside the unit circle if $|\alpha(i)| < 1$. Thus the fact that $\phi_2(z, i)$, $i = 1, 2, ..., n$, have all there zeros outside the closed unit disk if $|\alpha(i)| < 1$ for $i = 1, 2, ..., n$, follows by induction. That $w_1(z)$ and $w_2(z)$ are real while $w_3(z)$ and $w_4(z)$ are purely imaginary for z on the unit circle follows from the symmetry relations between $\phi^*(z, n)$ and $\phi(z, n)$, and $\phi_2(z, n)$ and $\phi_1(z, n)$.

If we examine

$$f(z) = \frac{(\phi^*(z, n) - \phi(z, n))}{(\phi(z, n) + \phi^*(z, n))} = \frac{1 - \frac{\phi(z, n)}{\phi^*(z, n)}}{1 + \frac{\phi(z, n)}{\phi^*(z, n)}},$$

then $\frac{\phi(z, n)}{\phi^*(z, n)}$ being a Schur function i.e. $|\frac{\phi(z, n)}{\phi^*(z, n)}| \leq 1$ for $|z| \leq 1$ implies that $f(z)$ is a C (Carathedory) function. Consequently $f(z)$ is a lossless rational of degree n (Delsarte, Genin [4]), i.e., $Re f(e^{i\theta}) = 0$ for almost all θ, which implies that the zeros of $z^{\frac{n}{2}} w_1(z)$ interlace those of $z^{\frac{n}{2}} w_3(z)$. An analogous argument gives the same result for the zeros of $z^{\frac{n}{2}} w_2(z)$ and $z^{\frac{n}{2}} w_4(z)$. To show that $w_1 w_2 - w_3 w_4 = 1$ we write

$$w_1 w_2 - w_3 w_4 = z^{-n} \frac{(\phi_1(z, n) \phi^*(z, n) - \phi_2(z, n) \phi(z, n))}{2} \tag{2.16}$$
$$= z^{-n} \det[\Phi_1(z, n), \Phi(z, n)] = 1,$$

where the Wronskian relation has been used to obtain the last equality. The above equation plus the fact that the zeros of $z^{\frac{n}{2}} w_1(z)$ interlace those of $z^{\frac{n}{2}} w_3(z)$ imply that between two consecutive zeros of w_1 lies a zero of w_4. Likewise the interlacing of the zeros of w_4 and w_2 coupled with (2.16) imply that between two consecutive zeros of w_4 lies a zero of w_1. An analogous argument shows that the zeros of $z^{\frac{n}{2}} w_2(z)$ and $z^{\frac{n}{2}} w_3(z)$ interlace which completes the lemma.

We now examine the eigenvalues of $T_{1,n}(z), \lambda_\pm$, which are given by

$$\lambda_\pm(z, n) = \frac{tr T_{1,n}(z) \pm \sqrt{(tr T_{1,n}(z))^2 - 4 \det T_{1,n}(z)}}{2}, \tag{2.17}$$

where the square is chosen so that $|x + \sqrt{x^2 - 1}| > 1$ for $x \in \mathbb{C} \setminus [-1, 1]$. If we define

$$Q^{\pm}(\theta, n) = w_1(e^{i\theta}) + w_2(e^{i\theta}) \pm 2, \qquad (2.18)$$

which is a real-valued function, the discriminant in the above equation can be written as $\sqrt{e^{in\theta}Q^+Q^-}$. At a zero of w_3 it follows from (2.15) that $Q^{\pm} = w_1 + \frac{1}{w_1} \pm 2$. Since the zeros of $z^{\frac{n}{2}} w_1(z)$ and $z^{\frac{n}{2}} w_3(z)$ interlace and $|x + \frac{1}{x}| \geq 2$ for x real we see that Q^+ and Q^- each change sign at most n times. Let $E = \{e^{i\theta} : |w_1(e^{i\theta}) + w_2(e^{i\theta})| \leq 2\}$, then from above we see that E may be composed of at most n subarcs of the unit circle. Let C_∞ be the component of the complement of E containing infinity.

THEOREM 2.2. *If $|\alpha(i)| < 1$ for $i = 1, 2, ..., n$, then $\frac{1}{n} \log |\lambda_+(z, n)| > 0$ for $z \in C_\infty$ and $\frac{1}{n} \log |\lambda_+(z, n)| = 0$ for $z \in E$. If E is the unit circle then $\frac{1}{n} \log |\lambda_-(z, n)| = 0$ for $|z| \leq 1$.*

Remark. Since $|\det T_{1,n}(z)| \geq 1$ when $|z| \geq 1$ at least one of the eigenvalues of $T_{1,n}$ is in magnitude greater than or equal to one for z in this region. It is the content of the theorem below that this is true even for $|z| < 0$.

Proof. For $z \in E$

$$\frac{1}{n} \log |\lambda_+(e^{i\theta}, n)|$$

$$= \frac{1}{n} \log \left| \frac{w_1(e^{i\theta}) + w_2(e^{i\theta}) + \sqrt{(w_1(e^{i\theta}) + w_2(e^{i\theta}))^2 - 4}}{2} \right| = 0,$$

because $|w_1(e^{i\theta}) + w_2(e^{i\theta})| \leq 2$. Also $\lim_{z \to \infty} (\frac{1}{n} \log |\lambda_+(z, n)| - \log |z|) = 0$. Since $\frac{1}{n} \log |\lambda_+(z, n)|$ is harmonic in C_∞ except at infinity the minimum principal for harmonic functions implies that $\frac{1}{n} \log |\lambda_+(z, n)| > 0$ for $z \in C_\infty$. The above argument shows that $\frac{1}{n} \log |\lambda_+(z, n)|$ is the Green's function associated with E, consequently $\frac{1}{n} \log |\lambda_+(z, n)| = \frac{1}{n} \log |z|$ if E is the unit circle. This implies that $|\lambda_-(z, n)| = 1$ for $|z| \geq 1$, since $\det T_{1,n} = z^n$. By examining λ_- we see that in this case $|\lambda_-| = 1$ also for $|z| < 0$ which completes the result.

We now examine ratios of of sums of solutions of (2.2). These will play an important role in the results that follow. In what follows we shall denote U as the unit circle.

LEMMA 2.3. *For each compact subset K of $\mathbb{C} \setminus \{U\}$ there exists a d such that*

$$\left| \frac{\phi(z, n) - \phi * (z, n)}{\phi(z, n) + \phi * (z, n)} \right| < d(1 - |\alpha(n)|)^{-2}, \qquad \left| \frac{\phi_1(z, n) + \phi_2(z, n)}{\phi(z, n) + \phi * (z, n)} \right| < d,$$

$$\left| \frac{\phi_1(z,n) - \phi_2(z,n)}{\phi_1(z,n) + \phi_2(z,n)} \right| < d(1 - |\alpha(n)|)^{-2} \quad z \in \hat{K}.$$

Proof. Let $\hat{K}$ be the part of K inside the unit circle. Since

$$s_n(z) = \frac{\phi * (z,n) - \phi(z,n)}{\phi(z,n) + \phi * (z,n)} \tag{2.20}$$

is a C-function it can be represented as (Akhieser [1, p. 91])

$$s_n(z) = iv_n + \int_{-\pi}^{\pi} \frac{e^{i\theta} + z}{e^{i\theta} - z} d\rho_n(\theta) \quad |z| < 1 \tag{2.21}$$

where ρ_n is a finite positive measure. With z equal to zero we find that

$$s_n(0) = \frac{1 - \alpha(n)}{1 + \alpha(n)} = iv_n + c_n$$

where $c_n = \int_{-\pi}^{\pi} d\rho_n(\theta)$. Therefore

$$c_n = \frac{1 - |\alpha(n)|^2}{|1 + \alpha(n)|^2} \leq \frac{1 + |\alpha(n)|}{1 - |\alpha(n)|}$$

and

$$|v_n| = \frac{|\bar{\alpha}(n) - \alpha(n)|}{|1 + \alpha(n)|^2} \leq \frac{2}{1 - |\alpha(n)|}.$$

Thus we find

$$|s_n(0)| \leq |v_n| + \int_{-\pi}^{\pi} \left| \frac{e^{i\theta} + z}{e^{i\theta} - z} \right| d\rho_n(\theta) \leq |v_n| + \hat{d} c_n$$

where $\hat{d}$ depends only upon the set $\hat{K}$. Combining the above inequalities we find $|s_n(0)| \leq c(1 - |\alpha(n)|)^{-2}$. The same reasoning holds for $\frac{\phi_2(z,n) - \phi_1(z,n)}{\phi_2(z,n) + \phi_1(z,n)}$. As for

$$g_n(z) = -\frac{\phi_2(z,n) + \phi_1(z,n)}{\phi * (z,n) + \phi(z,n)},$$

note that $Reg_n(e^{i\theta}) = 0$ for almost all θ. Furthermore by Lemma 1 the zeros of the numerator polynomial in g_n interlace those of the denominator polynomial, which implies that g_n is also a C-function with $g_n(0) = 1$. Therefore $g_n(z) \leq \hat{d}$ for $z \in \hat{K}$. For the part of K exterior to the unit disk $\tilde{K}, -s_n(z)$ is a C-function and has the representation

$$-s_n(z) = i\tilde{v}_n + \int_{-\pi}^{\pi} \frac{e^{i\theta} + z}{e^{i\theta} - z} d\tilde{\rho}_n(\theta) \quad |z| > 1.$$

From (2.20) we see that $-s_n(\infty) = \frac{1-\overline{\alpha(n)}}{1+\alpha(n)}$ which leads to the inequality $|s_n(z)| \leq \tilde{c}(1 - |\alpha(n)|)^{-2}$ where $\tilde{c}$ depends on $\tilde{K}$ but is independent of n. Analogous reasoning applies to $\frac{\phi_1(z,n)-\phi_2(z,n)}{\phi_2(z,n)+\phi_1(z,n)}$ and $-g_n$ for $z \in \tilde{K}$. Combining the inequalities for $z \in \hat{K}$ and $z \in \tilde{K}$ gives the result.

Consider now the solution of (2.1) given by

$$\Phi_+(z,n) = \begin{pmatrix} \phi_+^1(z,n) \\ \phi_+^2(z,n) \end{pmatrix} = \frac{1}{2}[\Phi_1(z,n) + F(z)\Phi(z,n)], \qquad |z| < 1 \quad (2.22)$$

with

$$F(z) = \int_{-\pi}^{\pi} \frac{e^{i\theta}+z}{e^{i\theta}-z}\, d\sigma. \quad (2.23)$$

THEOREM 2.4. *The sequence $\{\Phi_+(z,n)\}_{n=0}^{\infty}$ is the unique (up to a constant) l^2 solution of (2.1). $\Phi_+(z,n)$ has the representation*

$$\Phi_+(z,n) = \frac{1}{2}\int_{-\pi}^{\pi} \frac{e^{i\theta}+z}{e^{i\theta}-z}\, \Phi(e^{i\theta},n)d\sigma + \frac{1}{2}\begin{pmatrix} \delta_{0,n} \\ -\prod_{i=1}^{n}\frac{1}{a(i)} \end{pmatrix}, \quad (2.24)$$

and $\phi_+^1(z,n)$ is nonzero and greater in magnitude than $\phi_+^2(z,n)$ for $0 < |z| < 1$ and $n \geq 0$.

Proof. To prove (2.24) it suffices to show that

$$\Phi_1(z,n) = 2\Phi_1(z,n) - F(z)\Phi(z,n)$$

$$= \int_{-\pi}^{\pi} \frac{e^{i\theta}+z}{e^{i\theta}-z} (\Phi(e^{i\theta},n) - \Phi(z,n))d\sigma + \begin{pmatrix} \delta_{0,n} \\ -\prod_{i=1}^{n}\frac{1}{a(i)} \end{pmatrix}. \quad (2.25)$$

Apply $T(z,n+1)$ to both sides of the above equation and observe from (2.1) that

$$\Phi_1(z,n+1) = \int_{-\pi}^{\pi} \frac{e^{i\theta}+z}{e^{i\theta}-z}(\Phi(e^{i\theta},n+1) - \Phi(z,n+1))d\sigma$$

$$+ \int_{-\pi}^{\pi} \frac{e^{i\theta}+z}{e^{i\theta}-z}(T(z,n+1) - T(e^{i\theta},n+1))\Phi(e^{i\theta},n) \quad (2.26)$$

$$+ T(z,n+1)\begin{pmatrix} 0 \\ -\prod_{i=1}^{n}\frac{1}{a(i)} \end{pmatrix}$$

for $n \geq 1$. Since $T(z, n+1) - T(e^{i\theta}, n+1) = a(n+1)\begin{pmatrix} \dfrac{z - e^{i\theta}}{\alpha(n+1)(z - e^{i\theta})} & 0 \\ & 0 \end{pmatrix}$
the second integral on the right hand side becomes equal to

$$-\left(\frac{1}{\overline{\alpha}(n+1)}\right)\int_{-\pi}^{\pi} a(n+1)(e^{i\theta} + z)\phi(e^{i\theta}, n)d\sigma(\theta)$$

$$= a(n+1)\begin{pmatrix} \alpha(n+1) \\ |\alpha(n+1)|^2 \end{pmatrix}\prod_{i=1}^{n}\frac{1}{a(i)},$$

where (2.1) and orthonormality of $\phi(e^{i\theta}, n)$ has been used. This when added to
$T(z, n+1)\left(\begin{smallmatrix} 0 \\ -\prod_{i=1}^{n}\frac{1}{a(i)} \end{smallmatrix}\right)$ gives $\left(\begin{smallmatrix} 0 \\ -\prod_{i=1}^{n+1}\frac{1}{a(i)} \end{smallmatrix}\right)$ which when substituted into (2.26)
shows that both sides of (2.25) satisfy (2.1) with the same initial condition.
Thus (2.24) follows. Equation (2.24) now implies by the orthogonality of
$\phi(e^{i\theta}, n)$ that $\Phi_+(z, n) = \left(\begin{smallmatrix} 0(z^n) \\ 0(z^{n+1}) \end{smallmatrix}\right)$, thus $\{\Phi_+(z, n)\} \in l^2$ for $|z| < 1$. To
show that $\Phi_+(z, n)$ is unique up to a constant let $\{\Psi(z, n)\}$ be any other l^2
sequence then

$$W[\Psi(z, n), \Phi_+(z, n)] = \frac{1}{z^m}W[\Psi(z, n+m), \Phi_+(z, n+m)].$$

From the decay of $\Phi_+(z, n)$ it follows that the LHS of the above equation
tends to zero as m tends to infinity which in turn implies that $\{\Phi(z, n)\}$ and
$\{\Phi_+(z, n)\}$ are constant multiples of each other.

If we invert (2.1) we find

$$\frac{a(n)}{z}(\phi_+^1(z, n+1) - \alpha(n)\phi_+^2(z, n+1)) = \phi_+^1(z, n) \qquad (2.27)$$

and

$$a(n)(\phi_+^2(z, n+1) - \overline{\alpha(n)}\phi_+^1(z, n+1)) = \phi_+^2(z, n). \qquad (2.28)$$

Computing the square magnitude of both sides of the above equations
then subtracting the resulting equations yields

$$|\phi_+^1(z, n+1)^2 - |\phi_+^2(z, n+1)|^2 = |z|^2|\phi_+^1(z, n)|^2 - |\phi_+^2(z, n)|^2. \qquad (2.29)$$

This leads to the formula

$$|\phi_+^1(z, n+1)|^2 - |\phi_+^2(z, n+1)|^2$$
$$= |\phi_+^1(z, 0)|^2 - |\phi_+^2(z, 0)|^2 - (1 - |z|^2)\sum_{i=0}^{n}|\phi_+^1(z, i)|^2. \qquad (2.30)$$

Since $\Phi_+ \in l^2$

$$|\phi_+^1(z,0)|^2 - |\phi_+(z,0)|^2 = (1 - |z|^2) \sum_{i=0}^{\infty} |\phi_+^1(z,i)|^2, \qquad (2.31)$$

which implies that

$$|\phi_+^1(z,n)| \geq |\phi_+^2(z,n)|$$

for $|z| < 1$. If $|\phi_+^1(z,n)| = |\phi_+^2(z,n)|$ for $0 < |z| < 1$ then (2.30) and (2.31) imply that $\phi_+^1(z,i) = 0$ for $i \geq n$. Equation (2.1) now implies that $\phi_+^2(z,n) = 0$ for $i \geq n$ which is not possible by (2.22) and the linear independence of Φ_1 and Φ. Thus

$$|\phi_+^1(z,n)| > |\phi_+^2(z,n)|$$

for $|z| < 1$, and $n = 1, 2, \ldots$. From representation (2.24) we find that $\phi_+^1(z,0) - \phi_+^2(z,0) = 1$ which leads to the useful relation

$$|\phi_+^1(z,0)|^2 - |\phi_+^2(z,0)|^2 = 2 \operatorname{Re} \phi_+^1(z,0) - 1 = \int_{-\pi}^{\pi} \frac{1 - |z|^2}{|e^{i\theta} - z|^2} d\sigma. \qquad (2.32)$$

LEMMA 2.5. *Let*

$$m_+^1(z) = \frac{\phi_+^1(z,1)}{\phi_+^1(z,0)} \qquad (2.33)$$

and

$$m_+^2(z) = \frac{\phi_+^2(z,0)}{\phi_+^1(z,0)}. \qquad (2.34)$$

Then $\frac{m_+^1(z)}{z}$ and $\frac{m_+^2(z)}{z}$ are elements of H^∞ and $\frac{\log m_+^1(z)}{z}$ is in H^1.

Proof. Since $\phi_+^1(z,0)$ is not equal to zero for $0 \leq |z| < 1$, while (2.24) shows that $\phi_+^2(z,0) = 0$, $\frac{m_+^1(z)}{z}$ and $\frac{m_+^2(z)}{z}$ are analytic for $|z| < 1$. From (2.1) we find

$$\frac{m_+^1(z)}{z} = a(1) \left(1 + \alpha(1) \frac{m_+^2(z)}{z}\right). \qquad (2.35)$$

Theorem 2.4 and the Schwarz lemma tells us that $\left|\frac{m_+^2(z)}{z}\right| < 1$ for $|z| < 1$ since $|\alpha(1)| < 1$. Consequently $\frac{m_+^2(z)}{z}$ is in H^∞ and so is $\frac{m_+^1(z)}{z}$ from (2.35), (2.3) and the hypothesis that $|\alpha(1)| < 1$. Since $\left|\frac{m_+^1(z)}{z}\right| > 0$ for $|z| < 1$, $\log \frac{m_+^1(z)}{z}$ is analytic there hence $\log \frac{m_+^1(z)}{z}$ is in H^1 which completes the proof.

3. Lyapunov Exponent

We now assume that $\alpha(n)$ can be written in terms of a stationary stochastic process. More precisely let τ be an endomorphism of a finite measure space $(\Omega, \mathcal{A}, \mu)$ i.e $\tau : \Omega \to \Omega$ and $u(\tau^{-1}A) = \mu(A)$ for any $A \in \mathcal{A}$. Furthermore we will assume that τ is ergodic which means $\mu(A)$ equals zero or one for any $A \in \mathcal{A}$ such that $\tau^{-1}A = A$. Let f be a complex valued measurable function on Ω with the property that $|f(\Omega)| \subset [0,1)$. We note that since $\mu(\Omega)$ is finite we can without loss of generality assume $\mu(\Omega) = 1$. Define $\alpha_\omega(n) = f(\tau^n \omega)$ then,

$$T_n(\omega, z) = T_{n-1}(\tau\omega, z), \qquad\qquad z \in \mathbb{C} \qquad\qquad (3.1)$$

where $T_n(\omega, z)$ is given by (2.2). From now on we assume that

$$\mathbb{E}(\log(1 - |\alpha_\omega(0)|)) = \int \log(1 - |\alpha_\omega(0)|)d\mu(\omega) > -\infty \qquad (3.2)$$

which implies that $\log(1 - |\alpha_\omega(0)|) \in L(\mu)$. Consequently $\log(1 - |\alpha_\omega(n)|)/n$ tends to zero as n tends to infinity for μ – almost every ω by the ergodic theorem. We now set

$$\gamma_n(\omega, z) = \frac{1}{n} \log \| T_{1,n}(\omega, z) \| . \qquad\qquad (3.3)$$

THEOREM 3.1. *Given the above hypothesis on τ,*

$$\lim_{n \to \infty} \gamma_n(\omega, z) = \lim_{n \to \infty} 1/n \log \| T_{0,n}(\omega, z) \| = \gamma(z), \qquad (3.4)$$

exists and is independent of ω for each fixed $z \in \mathbb{C}$ and μ almost every ω. Furthermore $\gamma(z)$ is subharmonic, greater than or equal to zero, and for μ – almost every ω and every z, $\overline{\lim}_{n\to\infty}\gamma_n(\omega, z) \leq \gamma(z)$.

Remark. For convenience we will use the Hilbert–Schmidt norm.

Proof. Since $\|T_{1,1}(\omega, z)\| = a_\omega(1)[(1 + |z|^2)(1 + |\alpha_\omega|^2)]^{\frac{1}{2}}$ it follows from (2.3) and (3.2) that $\int \log^+ \|T_{0,1}(\omega, z)\|d\mu(\omega) < \infty$. Furthermore we find from (3.3) that $\log \|T_{i,j}\|$ is a subadditive stochastic process i.e., $\log \|T_{i,j} \circ \tau\| = \log \|T_{i+1,j+1}\|$, and $\log \|T_{i,j}\| \leq \log \|T_{i,k}\| + \log \|T_{k,j}\|, i \leq k \leq j$, consequently the existence of $\gamma(z)$ follows from the Kingman subadditive ergodic theorem, (Krengel [10]), which further asserts that $\gamma(z) = \lim_{n\to\infty} \int \gamma_n(\omega, z)d\mu = \inf_n \int \gamma_n(\omega, z)d\mu$. The fact that $\gamma(z) \geq 1$ is a consequence of Theorem 2.1 since at least one of the eigenvalues of $T_{1,n}(\omega, z)$ is in magnitude greater than or equal to one. The proof that $\gamma(z)$ is subharmonic is the same as the one given in Craig and Simon [2],(see also Herman [8]) and is based upon the fact that $\epsilon_n(z) = \int \log \|T_{1,n}(\omega, z)\| d\mu(\omega)$ is subharmonic, Rosenblum and Rovnyak [12, p. 75], and subadditive. This implies that $\frac{\epsilon_{2n}(z)}{2^n}, n = 0, 1, 2, \ldots$ is a decreasing sequence of subharmonic functions which converge to $\gamma(z)$. That $\overline{\lim}_{n \to \infty} \gamma_n|\omega, z| \leq \gamma(z)$ follows exactly as in Theorem 2.3 of Craig and Simon[2].

We now develop an integral relation between γ and a certain measure supported on the unit circle. In the case of polynomials orthogonal on the real line this measure is the so-called density of states measure (Simon [15], Geronimo, Harrell, Van Assche [5]). We begin by examining $\|T_{1,n}\|$ then using equations (2.12) and (2.13) which give

$$\|T_{1,n}\|^2 = \frac{1}{4}[|\phi(z, n) + \phi_1(z, n)|^2 + |\phi * (z, n) + \phi_2(z, n)|^2$$
$$+ |\phi(z, n) - \phi_1(z, n)|^2 + |\phi * (z, n) - \phi_2(z, n)|^2].$$

This in turn can be rewritten as

$$\|T_{1,n}\|^2 = \frac{1}{2}|\phi(z, n) + \phi * (z, n)|^2 \left[1 + \left|\frac{\phi(z, n) - \phi * (z, n)}{\phi(z, n) + \phi * (z, n)}\right|^2 \right.$$
$$\left. + \left|\frac{\phi_1(z, n) + \phi_2(z, n)}{\phi(z, n) + \phi * (z, n)}\right|^2 \left(1 + \left|\frac{\phi_1(z, n) - \phi_2(z, n)}{\phi_1(z, n) + \phi_2(z, n)}\right|^2\right)\right]. \tag{3.5}$$

If we take the log of the first term on the left hand side of (3.5) we find that

$$\frac{1}{n} \log \left|\frac{\phi(z, n) + \phi * (z, n)}{2}\right| = \frac{1}{n} \log \left|\frac{1 + \alpha(n)}{2 \prod_1^n a(i)}\right| + \int_{-\pi}^{\pi} \log |z - e^{i\theta}| d\beta_n(\theta), \tag{3.6}$$

where from Lemma 1.1 β_n is an atomic measure whose atoms are of size $\frac{1}{n}$ located at the zeros of $\phi(z, n) + \phi * (z, n)$ which are on the unit circle.

THEOREM 3.2. For μ almost all ω

$$\lim_{n \to \infty} \frac{-1}{n} \log \prod_1^n a(i) + \int_{-\pi}^{\pi} \log |z - e^{i\theta}| d\beta_n(\theta)$$
$$= \gamma(z) = R + \int_{-\pi}^{\pi} \log |z - e^{i\theta}| d\beta(\theta), \tag{3.7}$$

uniformly on compact subsets of $\mathbb{C} \setminus U$. Here $\lim_{n \to \infty} \frac{-1}{n} \log \prod_1^n a(i) = R, |R| < \infty$ and β is the weak limit of β_n. If we define $\int_{-\pi}^{\pi} \log |z - e^{i\theta}| d\beta(\theta)$, *by the convention that it is $-\infty$ if the integral liverges to $-\infty$ then the second equality in (3.7) holds for all z.*

Proof. Let K be a compact subset of $\mathbb{C} \setminus U$ then Lemma 2.3 shows that there exists positive constants c and c_1 which depend upon K but are independent of n such that

$$\left| \gamma_n(z, \omega) - \frac{1}{n} \log \left| \frac{\phi(z, n) + \phi * (z, n)}{2} \right| \right| \le \frac{c}{n} \max(\log 3, -\log c_1 (1 - |\alpha_\omega(n)|)).$$

Letting n tend to infinity and using Theorem 2.1 gives the first equality in (3.7). That $R = \lim_{n \to \infty} \frac{-1}{n} \log \prod_1^n a(i)$, follows from (3.7) by setting z equal to zero. The convergence of $\int_{-\pi}^{\pi} \log |z - e^{i\theta}| \, d\beta_n(\theta)$ to $\int_{-\pi}^{\pi} \log |z - e^{i\theta}| \, d\beta(\theta)$ for all $z \notin \mathbb{C} \setminus U$ implies that β_n converges weakly to β. Finally that the second equality holds for all z follows from that fact that $\gamma(z)$ and $R + \int_{-\pi}^{\pi} \log |z - e^{i\theta}| \, d\beta(\theta)$ are both subharmonic and equal to each other for almost all z.

Corollary 3.3. *For $|z| > 1$ and for μ almost all ω*

$$\lim_{n \to \infty} \frac{1}{n} \ln |\phi(z, n)| = \gamma(z),$$

where the convergence is uniform on compact subsets of $\mathbb{C} \setminus$ closed unit disk.

Proof. Since $|\phi(z, n)| > 0$ for $|z| > 1$ for all n, $\left| \frac{\phi^*(z, n)}{\phi(z, n)} \right| < 1$ for all n. The same is true for $\phi_1(z, n)$ and $\frac{\phi_2(z, n)}{\phi_1(z, n)}$ from Lemma 2.1. Therefore

$$\|T_{1,n}\|^2 = |\phi(z, n)|^2 \left[\left| \frac{\phi^2(z, n)}{\phi(z, n)} \right|^2 + \left| \frac{\phi_1(z, n)}{\phi(z, n)} \right|^2 \left[1 + \left| \frac{\phi_2(z, n)}{\phi_1(z, n)} \right|^2 \right] \right]$$

and the result follows since $\left\{ \frac{\phi_1(z, n)}{\phi(z, n)} \right\}$ converges to $\int \frac{e^{i\theta} + z}{e^{i\theta} - z} \, d\sigma$ uniformly on compact subsets of $\mathbb{C} \setminus$ closed unit disk.

4. m–Function

Set

$$m_+^1(z, \omega) = \frac{\phi_+^1(z, 1, \omega)}{\phi_+^1(z, 0, \omega)} \tag{4.1}$$

and

$$m_+^2(z, \omega) = \frac{\phi_+^2(z, 0, \omega)}{\phi_+^1(z, 0, \omega)}. \tag{4.2}$$

Since $\Phi_+(z, n, \tau^m\omega)$ and $\Phi_+(z, n + m, \omega)$ satisfy the same equation and are both in l^2 we find from Theorem 2.4 that $\Phi_+(z, n, \tau^m\omega) = c(z)\Phi_+(z, n+m, \omega)$ where $c(z)$ is independent of n. This implies that

$$m_+^1(z, \tau^m\omega) = \frac{\phi_+^1(z, m + 1, \omega)}{\phi_+^1(z, m, \omega)} \tag{4.3}$$

and

$$m_+^2(z, \tau^m\omega) = \frac{\phi_+^2(z, m, \omega)}{\phi_+^1(z, m, \omega)}. \tag{4.4}$$

THEOREM 4.1. *Set*

$$g(z) = \mathbb{E}\left[\log\frac{m_+^1(z, \omega)}{z}\right] \tag{4.5}$$

Then $g(z)$ is in H^1 and $\mathrm{Re}\, g(z) = -\gamma(z)$.

Proof. It follows easily from (2.35) that

$$\mathbb{E}\left|\log\left|\frac{m_+^1(z, \omega)}{z}\right|\right| < \infty$$

where the above integral converges uniformly on compact subsets of the open unit disk. This implies that $g(z)$ is analytic for $|z| < 1$. If we integrate the magnitude of g around the circle of radius $|r| < 1$ and use Fubini's theorem we find

$$\lim_{r\to 1}\int_{-\pi}^{\pi}|g(re^{i\theta})|d\theta \le \lim_{r\to 1}\mathbb{E}\left(\int_{-\pi}^{\pi}\left|\log\frac{m_+^1(re^{i\theta}, \omega)}{re^{i\theta}}\right|d\theta\right).$$

The interior integral on the RHS of the above equation is a monotonically increasing function of r. Consequently Lemma 2.5 and the monotone convergence theorem imply

$$\lim_{r\to 1}\int_{-\pi}^{\pi}|g(re^{i\theta})|d\theta \le \mathbb{E}\left(\lim_{r\to 1}\int_{-\pi}^{\pi}\left|\log\frac{m_+^1(re^{i\theta}, \omega)}{re^{i\theta}}\right|d\theta\right). \tag{4.6}$$

From (2.35) we find

$$1 - |\alpha_\omega(1)| \le 1 - \left|\frac{\alpha_\omega(1)m_+^2(z, \omega)}{z}\right| \le \left|\frac{m_+^1(z, \omega)}{z}\right| \le \alpha_\omega(1)\ln 2. \tag{4.7}$$

Substituting (4.7) into (4.6) shows that g is in H^1.

Since $\phi_+(z,n) \in l^2$ and det $T(z,n) = z$ for $|z| < 1$ we find from the Osceledec Theorem (Ruelle [12], Krengel [10]) that

$$\lim_{n\to\infty} \frac{1}{n} \log \|\Phi_+(z,n)\| = \ln |z| - \gamma(z). \qquad (4.8)$$

Furthermore

$$\lim_{n\to\infty} \frac{1}{n} \log \|\Phi_+(z,n)\|$$

$$= \lim_{n\to\infty} \frac{1}{n} \log \left[|\phi_+^1(z,n)|^2 + |\phi_+^2(z,n)|^2 \right] = \lim_{n\to\infty} \frac{1}{n} \log \left| \frac{\phi_+^1(z,n)}{\phi_+^1(z,0)} \right|$$

where the boundedness $\left| \frac{\phi_+^2(z,n)}{\phi_+^1(z,n)} \right|$ was used to obtain the last inequality. Substitution of (4.3) into the above equation yields

$$\lim_{n\to\infty} \frac{1}{n} \log \left| \frac{\phi_+^1(z,n)}{\phi_+^1(z,0)} \right| = \lim_{n\to\infty} \frac{1}{n} \sum_{i=0}^{n-1} \log |m_+^1(z,\tau^n\omega)|.$$

Since $\log \left| \frac{m_+^1(z,\omega)}{z} \right| \in L^1(\mu)$ the ergodic theorem implies that Re $g(z) = -\gamma(z)$.

LEMMA 4.3.

$$\gamma(z) - \ln |z| = \mathbb{E}\left(\frac{1}{2} \log \left[1 + \frac{1 - |z|^2}{|z|^2 - |m_+^2(z,\omega)|^2} \right] \right), \qquad (4.9)$$

for $|z| < 1$.

Proof. From (2.28) with $n = 0$ we find using (4.3) and (4.4) that

$$|m_+^1(z,\omega)|^2 = \frac{|z|^2 - |m_+^2(z,\omega)|^2}{1 - |m_+^2(z,\tau\omega)|^2}. \qquad (4.10)$$

The fact that τ is an endomorphism implies that

$$-E[\ln |(m_+^1(z,\omega))|^2] = E[\log(1 - |m_+^2(z,\omega)|^2)] - E[\log(|z|^2 - |m_+^2(z,\omega)|^2)]$$

$$= E\left[\log \left[1 + \frac{1 - |z|^2}{|z|^2 - |m_+^2(z,\omega)} \right]^2 \right],$$

and the result now follows from Theorem 4.2.

LEMMA 4.4.

$$\gamma(z) - \ln|z| > \mathbb{E}\left\{\ln\left[1 + \frac{1 - |z|^2}{4(2\ \mathrm{Re}\ \phi_+^1(z, 0, \omega) - 1}\right]\right\}, \qquad (4.11)$$

for $|z| < 1$.

Proof. Equation (4.2) implies that

$$\frac{1 - |z|^2}{|z|^2 - |m_+^2(z, \omega)|}^2 = \frac{(1 - |z|^2)|\phi_+^1(z, 0, \omega)|^2}{|z|^2|\phi_+^1(z, 0, \omega)|^2 - |\phi_+^2(z, 0, \omega)|^2}$$

$$> \frac{(1 - |z|^2)|\phi_+^1(z, 0, \omega)|^2}{|\phi_+^1(z, 0, \omega)|^2 - |\phi_+^2(z, 0, \omega)|^2}.$$

Examination of (2.23) shows that $|\phi_+^1(z, 0, \omega)| > \mathrm{Re}\ \phi_+^1(z, 0, \omega) \geq 1/2$ which proves the lemma.

LEMMA 4.5. *For* μ – *almost every* ω *and* $0 < |z| < 1$

$$\log\frac{1 - |m_+^2(z, \omega)|^2}{|z|^2 - |m_+^2(z, \omega)|^2} \leq \log\frac{1 - |m_+^2(z, \tau^n\omega)|^2}{|z|^2 - |m_+^2(z, \tau^n\omega)|^2}$$

$$+ \log\left[1 + \frac{2(1 - |z|^2)(\frac{2}{|z|})^{2n+2}}{\left(\frac{4}{|z|^2} - 1\right)\left(|z|^2 - |m_+^2(z, \tau^n\omega)|^2\right)}\right]$$

Proof. From Eqs. (2.30) and (4.4) we find

$$1 - |m_+^2(z, \omega)|^2 =$$

$$= \left|\frac{\phi_+^1(z, n, \omega)}{\phi_+^1(z, 0, \omega)}\right|^2\left(1 - |m_+^2(z, \tau^n\omega)|^2 + (1 - |z|^2)\sum_{i=0}^{n-1}\left|\frac{\phi_+^1(z, i, \omega)}{\phi_+^1(z, n, \omega)}\right|^2\right),$$

and

$$|z|^2 - |m_+^2(z, \omega)|^2 =$$

$$= \left|\frac{\phi_+^1(z, n, \omega)}{\phi_+^1(z, 0, \omega)}\right|^2\left(|z|^2 - |m_+^2(z, \tau^n\omega)|^2 + (1 - |z|^2)\sum_{i=1}^{n}\left|\frac{\phi_+^1(z, i, \omega)}{\phi_+^1(z, n, \omega)}\right|^2\right).$$

Eq. (2.1) shows us that

$$\frac{\phi_+^1(z, n, \omega)}{\phi_+^1(z, i, \omega)} = \prod_{j=i}^{n-1}\frac{\phi_+^1(z, j + 1, \omega)}{\phi_+^1(z, j, \omega)} = \prod_{j=i}^{n-1}a_\omega(j)\left(z + \alpha_\omega(j)\frac{\phi_+^2(z, j, \omega)}{\phi_+^1(z, j, \omega)}\right).$$

Theorem 2.4 and the Schwarz lemma imply that for almost all ω, $\left|\frac{\phi^2_+(z,j,\omega)}{\phi^1_+(z,j,\omega)}\right| <$ $|z|$, consequently

$$\left|\frac{\phi^1_+(z,n,\omega)}{\phi^1_+(z,i,\omega)}\right| \geq |z|^{n-i} \prod_{j=i}^{n-1} a(j)[1 - |\alpha_\omega(j)|] = |z|^{n-i} \prod_{j=i}^{n-1} \frac{1}{1 + |\alpha_\omega(j)|}$$ (4.12)

$$\geq \frac{|z|^{n-i}}{2^{n-i}}$$

where (2.3) has been used to obtain the last inequality. Since

$$\log\frac{1 - |m^2_+(z,\omega)|^2}{|z|^2 - |m^2_+(z,\omega)|^2} \leq \log\frac{1 - |m^2_+(z,\tau^n\omega)|^2}{|z|^2 - |m^2_+(z,\omega)|^2}$$

$$+ \log\left[1 + (1 - |z|^2)\left|\frac{\sum_{i=0}^{n-1}\left|\frac{\phi^1_+(z,e,\omega)}{\phi^1_+(z,n,\omega)}\right|^2}{1 - |m^2_+(z,\tau^n\omega)|^2} - \frac{\sum_{i=1}^{n}\left|\frac{\phi^1_+(z,e,\omega)}{\phi^1_+(z,n,\tau^n\omega)}\right|^2}{|z|^2 - |m^2_+(z,\tau^n\omega)|^2}\right|\right]$$

the result now follows from (4.12) because $1 - |m^2_+(z)|^2 \leq |z|^2 - |m^2_+(z)|^2$ for $|z| < 1$.

Let $\sum_{a.c.}(\omega)$ be the support of the absolutely continuous components of σ^ω and denote the density function associated with this component by $\hat\sigma^\omega_{a.c.}$. We now prove the following:

THEOREM 4.6. *For almost every ω*

$$\sum_{a.c.}(\omega) = \{e^{i\theta} : \gamma(e^{i\theta}) = 0\},$$

up to subsets of Lebesgue measure zero.

Remark. The second half of this theorem draws upon some of the ideas in Deift and Simon [3].

Proof. Let $A = \{e^{i\theta} : \gamma(e^{i\theta}) = 0\}$ and suppose A has positive Lebesgue measure. From Theorems (3.2) and (4.2) we find that Re $(1 - zg'(z)) > 0$ for $|z| < 1$. This implies that $\log(1 - zg'(z)) \in H_2$ hence $\lim_{r\to1} g'(re^{i\theta})$ exists for almost every θ. Since

$$\frac{dg}{dz} = e^{-i\theta_0}\left(\frac{\partial\,\mathrm{Re}\,g}{\partial r} + i\frac{\partial\,\mathrm{Im}\,g}{\partial r}\right) \quad z = re^{-i\theta_0},$$

we find using Theorem (4.2) that $\lim_{r\to1} \frac{d\gamma(re^{i\theta_0})}{dr}$ exists for almost every θ_0. Furthermore for any such θ_0 where also $e^{i\theta_0} \in A$ we find

$$\lim_{r\to1} \frac{\gamma(re^{i\theta_0})}{1 - r} = \lim_{r\to1} -\gamma'(re^{i\theta_0}) < \infty.$$ (4.13)

Lemma 4.4 and (4.13) and the inequality $\log(1+x) \geq \frac{x}{1+x}$, $x \geq 0$ yield

$$\limsup_{\substack{r \to 1 \\ r < 1}} \mathbb{E}\left(\frac{1}{2 \operatorname{Re}\phi_+^1(re^{i\theta_0}, 0, \omega) - 1}\right) < \infty. \qquad (4.14)$$

The integral representation for $2\phi_+^2(z, 0, \omega) - 1$ arising from (2.32) tells us that $2\phi_+^1(z, 0, \omega) - 1$ has radial boundary values for all ω and almost all θ. Consequently by (4.13) and Fatou's lemma

$$\mathbb{E}\left(\frac{1}{2 \operatorname{Re} \phi_+^1(e^{i\theta_0}, 0, \omega) - 1}\right) < \infty.$$

Thus for μ – almost all ω and almost every $e^{i\theta} \in A$, $2 \operatorname{Re} \phi_+^1(e^{i\theta}, 0, \omega) - 1 > 0$. Since $\hat{\sigma}_{a.c.}^\omega(\theta) = [2 \operatorname{Re} \phi_+^1(e^{i\theta}, 0, \omega) - 1]$ one half of the theorem follows.

To show the other half let $A \subset U$ such that $|A| > 0$ and for μ – almost every ω $\hat{\sigma}_{a.c.}^\omega > 0$ for Lebesgue almost every $e^{i\theta} \in A$. There is a set $B \subset A$ with $|B| = |A|$ such that for μ – almost every ω and all $e^{i\theta} \in B$, $\hat{\sigma}_{a.c.}^\omega > 0$. This implies by redefining B if necessary that $\operatorname{Re} F(e^{i\theta}, \omega) > 0$ for almost all ω and $e^{i\theta} \in B$. Consequently

$$\lim_{r \to 1} \log \frac{1 - |m_+^2(re^{i\theta}, \omega)|^2}{|z|^2 - |m_+^2(re^{i\theta}, \omega)|^2}$$
$$= \lim_{r \to 1} \log \frac{4}{2(1 + r^2) - (1 - r^2)\frac{(|F(re^{i\theta}, \omega)|^2 + 1)}{\operatorname{Re} F(re^{i\theta}, \omega)}} = 0 \qquad (4.15)$$

and

$$\lim_{r \to 1} \frac{1 - r^2}{r^2 - |m_+^2(re^{i\theta}, \omega)|^2} = 0. \qquad (4.16)$$

for μ – almost every ω and every $e^{i\theta} \in B$ since $F(re^{i\theta}, \omega)$ has finite radial boundary values for every ω and almost every $e^{i\theta}$. By Ergoroff's theorem there exists a set $E \subset \Omega$ whose μ measure may be made arbitrarily close to one such that the convergence in Eqs. (4.15) and (4.16) is uniform on E. By the ergodic theorem there exists a set $\hat{E}^c \subset E^c$ so that $\mu(\hat{E}^c) = \mu(E^c)$ and there is an N_0 such that $\bigcup_{i=1}^{N_0} \tau^n \omega \cap E \neq 0$ for all $\omega \in \hat{E}^c$. Thus for any ω in $\hat{E}^c$ there exists an $\omega_0 \in E$ with the property that $\tau^n \omega = \omega_0$ for some $n \leq N_0$. Consequently Lemma 4.5 implies

$$\log \frac{1 - |m_+^2(z, \omega)|^2}{|z|^2 - |m_+^2(z, \omega)|^2} \leq \log \frac{1 - |m_+^2(z, \omega_0)|^2}{|z|^2 - |m_+^2(z, \omega_0)|^2}$$
$$+ \log\left[1 + \frac{1 - |z|^2 (2/|z|)^{2N_0 + 2}}{\left(\frac{4}{|z|^2} - 1\right)|z|^2 - |m_+^2(z, \omega_0)|^2}\right] \qquad (4.17)$$

Hence for $e^{i\theta} \in B$

$$
\begin{aligned}
\lim_{r \to 1} \gamma(re^{i\theta}) &= \lim_{r \to 1} \int \frac{1}{2} \log \frac{1 - |m_+^2(re^{i\theta}, \omega)|^2}{r^2 - |m_+^2(re^{i\theta}, \omega)|^2} d\mu(\omega) \\
&= \lim_{r \to 1} \int_E \frac{1}{2} \log \frac{1 - |m_+^2(re^{i\theta}, \omega)|^2}{r_2 - |m_+^2(re^{i\theta}, \omega)|^2} d\mu(\omega) \\
&\quad + \lim_{r \to 1} \int_{\hat{E}^c} \frac{1}{2} \log \frac{1 - |m_+^2(re^{i\theta}, \omega)|^2}{r^2 - |m_+^2(re^{i\theta}, \omega)|^2} d\mu(\omega)
\end{aligned}
\tag{4.18}
$$

That the right hand side of the above equation is equal to zero follows from the fact that the integrand in the first term converges uniformly to zero and (4.17).

5. Periodic Coefficients

The case when the probability space Ω is a finite number of points N gives rise to sequences $\{\alpha(n)\}_{n=1}^\infty$ which are periodic of period N. In this case

$$
T_{1,i} = T_{nN+1,i} T_{1,N}^n
$$

where $i = nN + j, 0 \le j < N$ and $T_{j,i} = 1$ for $j > i$. Therefore for $|z| > 1$,

$$
\gamma(z) = \frac{1}{N} \log \|\lambda_+(z, N)\|
$$

where $\lambda_+(z, N)$ is given by equation (2.17). From the proof of Theorem 2.2 we see that $\gamma(z)$ is the Green's function for the set E defined just before Theorem 2.2. This implies that R defined in Theorem 3.2 is $1/\log|C(E)|$ where $C(E)$ is the capacity of the set E. Furthermore $\beta(\theta)$ is the equilibrium measure associated with E and has the form

$$
\begin{aligned}
\frac{d\beta(\theta)}{d\theta} &= \lim_{r \to 1} \frac{1}{2\pi} \frac{\partial}{\partial r} \gamma(re^{i\theta}) \\
&= \frac{1}{2\pi} \left[\frac{1}{2} + \frac{1}{N} \frac{\operatorname{Im}\left\{ \frac{d}{dr} Tr((re^{i\theta})^{-\frac{N}{2}} T_{1,N}(re^{i\theta}))|_{r=1} \right\}}{\sqrt{4 - \left[Tr\left(e^{-iN\theta/2} T_{1,N}(e^{i\theta})\right) \right]^2}} \right], \quad e^{i\theta} \in E,
\end{aligned}
$$

where the fact that $Tr(e^{-iN\theta/2} T_{1,N}(e^{i\theta})) = \frac{w_1(e^{i\theta}) + w_2(e^{i\theta})}{2}$, which is real, has been used to obtain the last equation. The branch of the square root used in the above equation is the one so that $|x + \sqrt{x^2 - 1}| > 1$ for $x \in \mathbb{C} \setminus [-1, 1]$. From Theorem 4.6 we see that the absolutely continuous components of the N different orthogonality measures associated $\{\alpha(n)\}_{n=1}^\infty$ are all supported on E.

References

[1] Akhiezer N. I., *The Classical Moment Problem*, Hafner, N.Y., 1965.

[2] Craig W. and Simon B., *Subharmonicity of the Lyapunov index*, Duke Math. J. **50** (1983), pp. 551–560.

[3] Deift P. and Simon B., *Almost periodic Schrödinger operators III. The absolutely continuous spectrum in one dimension*, Commun. Math. Phys. **90** (1983), pp. 389–411.

[4] Delsarte P. and Genin Y., *The tridiagonal approach to Szegö's orthogonal polynomials, Toeplitz systems, and related interpolation problems*, SIAM J. Math. Anal. **19** (1988), pp. 718–738.

[5] Geronimo J. S., W., Harrell II E. M. and Van Assche W., *On the asymptotic distribution of eigenvalues of banded matrices*, Constr. Approx. **4** (1988), pp. 403–417.

[6] Geronimo J. S. and Johnson R., *Rotation numbers associated with difference equations satisfied by polynomials orthogonal on the unit circle*, in preparation..

[7] Geronimus Yu L., *Polynomials Orthogonal on a Circle and Interval,*, Pergamon Press, NY, 1960.

[8] Herman M., *Une methode pour minorer les exposants de Lyapounov et quelques exemples montrant le caractere local d'un theoreme d'Arnold et de Moser sur le tore de dimension 2*, Comment. Math. Helvetici **58** (1982), pp. 453–502.

[9] Kotani S., *Lyapunov indices determine absolutely continuous spectra of stationary random one-dimensional Schrödinger operators*, Proc. Taniguchi Symp. SA Katata, (1982), pp. 225–247.

[10] Krengel U., *Ergodic Theorems*, de Gruyter, NY, 1985.

[11] Nikishin E.M., *Random Orthogonal Polynomials on a circle*, Vestnik. Moskovskogo Universitita Matematika **42** (1987), pp. 52–55.

[12] Rosenblum M. and Rovnyak J., *Hardy Classes and Operator Theory*, Clarendon, NY, 1985.

[13] Ruelle D., *Characteristic exponents and invariant manifolds in Hilbert space*, Annal of Math. **115** (1982), pp. 243–290.

[14] Simon B., *Kotani theory for one dimensional stochastic Jacobi matrices,*, Commun. Math. Phys. **89** (1983), pp. 227–234.

[15] Simon B., *Schrödinger semigroups*, Bull. AMS **7** (1982), pp. 447–526.

[16] Teplyaev A.V., *Orthogonal polynomials with random recurrence coefficients*, LOMI Scientific Seminar notes **117** (1989).

[17] Szegö G., *Orthogonal polynomials*, AMS Coll. Pub. **23** (1978).

Math.Dept.,Georgia Tech.
Atlanta GA 30332
USA

USING THE REFINEMENT EQUATION FOR THE CONSTRUCTION OF PRE-WAVELETS IV: CUBE SPLINES AND ELLIPTIC SPLINES UNITED

Charles A. Micchelli

1. Introduction

Motivated by a multivariate *tensor product* pre-wavelet construction given by Battle [1], we provided in [2] a concrete pre-wavelet decomposition of L^2 using cube splines. Later this construction was extended to a general class of functions which include polyharmonic B-splines [4]. In this paper we focus on the essential aspects of [2] and [4] to obtain a theorem which unifies some of the conclusions in these papers. This also gives us the opportunity to provide an orthogonal pre-wavelet decomposition of L^2 relative to an arbitrary lattice of $\mathbb{Z}^s$ induced by a nonsingular $s \times s$ integer matrix whose inverse is contractive.

2. Pre-Wavelets

We suppose that $\varphi \in L^2(\mathbb{R}^s)$ and M is an $s \times s$ integer matrix with $n := |\det M| > 1$. We say that the pair (φ, M) admits multiresolution if the following properties hold:

(i) $V^k \subseteq V^{k+1}$, $k \in \mathbb{Z}$ where $V^k := sc^k V^\circ$, $(sc^k f)(x) := f(M^k x)$, $x \in \mathbb{R}^s$, $f \in L^2(\mathbb{R}^s)$, $V^\circ = V^\circ(\varphi) := \{[d, \varphi] : d \in \ell^2(\mathbb{Z}^s)\}$ and

$$[d, \varphi](x) := \sum_{\alpha \in \mathbb{Z}^s} d_\alpha \varphi(x - \alpha).$$

(ii)   There are constants $0 < m < k < \infty$ such that

$$m\|d\|_2 \leq \|[d, \varphi]\|_2 \leq k\|d\|_2$$

where $\|\cdot\|_2$ is the L^2 or ℓ^2 on $\mathbb{R}^s$, $\mathbb{Z}^s$ respectively, and

(iii)

$$\bigcup_{k \in \mathbb{Z}} V^k = L^2(\mathbb{R}^s), \quad \bigcap_{k \in \mathbb{Z}} V^k = \{0\}.$$

The work of this author has been partially supported by a DARPA grant.

Our first requirement (i) implies that φ satisfies a refinement equation

$$\varphi(x) = \sum_{\alpha \in \mathbf{Z}^s} a_\alpha \varphi(Mx - \alpha), \quad x \in \mathbb{R}^s \tag{2.1}$$

for some constants a_α, $\alpha \in \mathbf{Z}^s$. We assume throughout that $a = (a_\alpha : \alpha \in \mathbf{Z}^s) \in \ell^1(\mathbf{Z}^s)$.

For every $d \in \ell^1(\mathbf{Z}^s)$ we associate the Laurent series,

$$d(w) := \sum_{\alpha \in \mathbf{Z}^s} d_\alpha e^{i\alpha \cdot w}, \quad w \in \mathbb{R}^s.$$

Our basic hypothesis on $a = (a_\alpha : \alpha \in \mathbf{Z}^s)$ in (2.1) is that there is a $b \in \ell^1(\mathbf{Z}^s)$ with meas $\{w : b(w) = 0, \ w \in [-\pi, \pi]^s\} = 0$ and a $\rho \in \mathbb{C} \backslash \{0\}$ such that

$$\rho b(M^t w) = b(w)a(w), \quad w \in \mathbb{R}^s, \tag{2.2}$$

($M^t = $ transpose of M). In the next section we will provide examples where these hypotheses are satisfied.

The matrix M induces n sublattices of $\mathbb{R}^s$, that is, there are vectors $e^0, \ldots, e^{n-1} \in \mathbf{Z}^s$ with $e^0 = 0$ such that the sublattices

$$L_k := e^k + M\mathbf{Z}^s, \ 0 \leq k < n \tag{2.3}$$

form a partition of $\mathbf{Z}^s$, that is $\bigcup_{k=0}^{n-1} L_k = \mathbf{Z}^s$ and $L_k \cap L_{k'} = \phi$, if $k \neq k'$. Let $E = \{e^j : 1 \leq j \leq n-1\}$.

Now, for every $\psi \in V^1$ we associate the $n-1$ functions

$$\psi_e(x) := \psi(x - M^{-1}e), \quad e \in E \backslash \{0\}. \tag{2.4}$$

Each ψ_e generates a scale of spaces, *viz.*

$$W_e^k := sc^k V^o(\psi_e).$$

Following [5], we consider the subspace $\mathcal{L}^2(\mathbb{R}^s)$ consisting of all functions $\varphi \in L^2(\mathbb{R}^s)$ such that

$$\varphi^o(x) := \sum_{\alpha \in \mathbf{Z}^s} |\varphi(x - \alpha)|, \tag{2.5}$$

is in $L^2([0,1]^s)$

THEOREM 2.1. *Suppose $\varphi \in \mathcal{L}^2$ and the pair (φ, M) admits multiresolution with $\lim_{k \to \infty} M^{-k} = 0$. Let the coefficients $a = (a_\alpha : \alpha \in \mathbf{Z}^s)$ in the refinement equation (2.1) satisfy (2.2) for some $\rho \in \mathbb{C} \backslash \{0\}$ with $|\rho| \neq \sqrt{n}$ and $b \in \ell^1$ with meas $\{w : b(w) = 0, \ w \in [-\pi, \pi]^s\} = 0$. Then there exists a $\psi = sc[d, \varphi] \in V^1$ with $d \in \ell^1$ such that*

(a) $V^k \perp W_e^k, \quad e \in E \backslash \{0\}$
(b) $W_e^k \perp W_{e'}^{k'}, \quad k \neq k', \ k, k' \in \mathbf{Z}, \ e, e' \in E \backslash \{0\}$,
(c) $\overline{\sum_{k \in \mathbf{Z}} W^k} = L^2(\mathbb{R}^s)$

where

$$W^k := \sum_{e \in E \setminus \{0\}} W_e^k.$$

Through the course of the proof of this result we will identify a specific choice of $d \in \ell^1(\mathbf{Z}^s)$ for which the corresponding ψ satisfies the conclusions of the Theorem.

First, we show that (a), (b) follow from the fact that

$$V^\circ \perp W_e^\circ, \quad e \in E \setminus \{0\}. \tag{2.6}$$

To see this we suppose that (2.6) holds. Then (a) follows from the trivial observation that $f \perp g$ if and only if $sc^k f \perp sc^k g$, for any $f, g \in L^2(\mathbb{R}^s)$. As for (b) we suppose for definiteness that $k' > k$. It follows that

$$W_e^k \subseteq V^{k+1} \subseteq V^{k'}$$

and so $W_e^k \perp W_{e'}^{k'}$, since $V^{k'} \perp W_{e'}^{k'}$. Thus the validity of (a) and (b) hinges upon (2.6). To explore this condition we recall that in [5] it was observed that when $\varphi \in \mathcal{L}^2$ the sequence $g = (g_\alpha : \alpha \in \mathbf{Z}^s)$,

$$g_\alpha := \int_{\mathbb{R}^s} \varphi(x)\overline{\varphi}(x - \alpha)dx, \quad \alpha \in \mathbf{Z}^s,$$

is in $\ell^1(\mathbf{Z}^s)$, the associated Laurent series $g(w)$ (in view of (ii)) is strictly positive for all $w \in \mathbb{R}^s$, and by the Poisson summation formula

$$g(w) = \sum_{\alpha \in \mathbf{Z}^s} g_\alpha e^{i\alpha \cdot w} = \sum_{\alpha \in \mathbf{Z}^s} |\hat{\varphi}(w + 2\pi\alpha)|^2 \tag{2.7}$$

where $\hat{\varphi}(w)$ is the Fourier transform of φ,

$$\hat{\varphi}(w) = \int_{\mathbb{R}^s} e^{iw \cdot x} \varphi(x) dx.$$

Let $h \in \ell^1(\mathbf{Z}^s)$ and suppose that $d \in \ell^1(\mathbf{Z}^s)$ is chosen to satisfy the equation

$$d(w)g(w)\overline{a(w)} = h(M^t w), \quad w \in \mathbb{R}^s. \tag{2.8}$$

Then

$$\int_{\mathbb{R}^s} \overline{\varphi}(x + \alpha)\psi_e(x)dx = (2\pi)^{-s} \int_{\mathbb{R}^s} e^{i\alpha \cdot w} \overline{\hat{\varphi}}(w)\hat{\psi}_e(w)dw$$

$$= (2\pi)^{-s} \int_{\mathbb{R}^s} e^{i(M\alpha + e) \cdot w} \overline{\hat{\varphi}}(M^t w)d(w)\hat{\varphi}(w)dw$$

$$= n^{-1}(2\pi)^{-s} \int_{\mathbb{R}^s} e^{i(M\alpha + e) \cdot w} |\hat{\varphi}(w)|^2 \frac{h(M^t w)}{g(w)}dw$$

$$= n^{-1}(2\pi)^{-s} \int_{[-\pi,\pi]^s} e^{i(M\alpha + e) \cdot w} h(M^t w)dw = 0,$$

since $e \in E \setminus \{0\}$.

As a specific choice of d we set

$$d(w) := \overline{b(w)}/g(w), \tag{2.9}$$

or, equivalently, in view of (2.2)

$$\overline{h(w)} := \rho b(w). \tag{2.10}$$

By Wiener's lemma [cf. W. Rudin [6], p.266] we see that $d \in \ell^1$ and therefore for this choice of d both (a) and (b) hold.

The last claim (c) of Theorem 2.1 is more complicated to prove. In this regard we use the following lemma.

LEMMA 2.1. *Let φ, ψ be as above. Then there is a $\gamma = (\gamma_\alpha : \alpha \in \mathbf{Z}^s) \in \ell^1(\mathbf{Z}^s)$ such that*

$$\psi(x) = \sum_{\alpha \in \mathbf{Z}^s} \gamma_\alpha \psi(Mx - M^{-1}\alpha) \tag{2.11}$$

with

$$\gamma_{M\alpha} = n\bar{\rho}^{-1}\delta_{o\alpha}, \quad \alpha \in \mathbf{Z}^s. \tag{2.12}$$

Equation (2.11) is a refinement equation (2.1) for $\Gamma := sc^{-1}\psi$ with the special property (2.12). It is interesting to note that this type of refinement equation appeared in [3] in the univariate case when $Mx = 2x$, $x \in \mathbb{R}$.

Proof: Let's identify γ working backwards from (2.11). Since $n\hat{\psi}(M^t w) = \gamma(M^{-t}w)\hat{\psi}(w)$ we have

$$d(w)\hat{\varphi}(w) = n^{-1}\gamma(M^{-t}w)d(M^{-t}w)\hat{\varphi}(M^{-t}w)$$

or

$$d(w)a(M^{-t}w) = \gamma(M^{-t}w)d(M^{-t}w).$$

We now use this last equation to define γ as

$$\gamma(w) := \frac{d(M^t w)}{d(w)}a(w) = \frac{\bar{b}(M^t w)}{\bar{b}(w)}\frac{g(w)}{g(M^t w)}a(w) = \bar{\rho}^{-1}\frac{|a(w)|^2 g(w)}{g(M^t w)}.$$

With this definition of γ it remains to check that (2.12) holds since obviously $\gamma \in \ell^1(\mathbf{Z}^s)$. To this end, we observe that for $\alpha \in \mathbf{Z}^s$

$$(2\pi)^{-s}\int_{[-\pi,\pi]^s} e^{iM\alpha \cdot w}\gamma(w)dw = \bar{\rho}^{-1}(2\pi)^{-s}\int_{\mathbb{R}^s} e^{iM\alpha \cdot w}\frac{|\hat{\varphi}(w)|^2|a(w)|^2}{g(M^t w)}dw$$

$$n^2\bar{\rho}^{-1}(2\pi)^{-s}\int_{\mathbb{R}^s} e^{iM\alpha \cdot w}\frac{|\hat{\varphi}(M^t w)|^2}{g(M^t w)}dw$$

$$= n\bar{\rho}^{-1}(2\pi)^{-s}\int_{[-\pi,\pi]^s} e^{i\alpha \cdot w}dw = n\bar{\rho}^{-1}\delta_{o\alpha}.$$

This proves the lemma.

COROLLARY 2.1. *Let ψ be as in Lemma 2.1 and set*

$$\Gamma = sc^{-1}\psi. \tag{2.13}$$

Suppose further that $\lim_{k\to\infty} M^{-k} = 0$ and $|\rho| \neq \sqrt{n}$. Then

$$\sum_{k\in\mathbf{Z}} V^k(\Gamma) \subseteq \overline{\sum_{k\in\mathbf{Z}} W^k}. \tag{2.14}$$

Proof. We let $\mathcal{B}$ be the set of all lattice points $\beta \in \mathbf{Z}^s$ such that for all $r \in \mathbf{Z}$

$$\Gamma(M^r \bullet -\beta) \in W := \sum_{k\in\mathbf{Z}} W^k.$$

We will show that $\mathcal{B} = \mathbf{Z}^s \backslash \{0\}$. To this end, we note that if $\beta = M\alpha + e$, $e \in E\backslash\{0\}$ then

$$\Gamma(M^r x - \beta) = \psi(M^{-1}(M^r x - M\alpha - e))$$
$$= \psi(M^{r-1}x - \alpha - M^{-1}e) = \psi_e(M^{r-1}x - \alpha).$$

Hence by definition of the space W^k we see that $M\alpha + e \in \mathcal{B}$ for all $\alpha \in \mathbf{Z}^s$ and $e \in E\backslash\{0\}$. Next, according to Lemma 2.1 we have

$$\Gamma(M^r x - \beta) = \sum_{\alpha\in\mathbf{Z}^s} \gamma_\alpha \psi(M^r x - \beta - M^{-1}\alpha)$$

$$= \sum_{e\in E} \sum_{\mu\in\mathbf{Z}^s} \gamma_{M\mu+e} \psi(M^r x - \beta - M^{-1}(M\mu + e))$$

$$= n\bar{\rho}^{-1}\Gamma(M^{r+1}x - M\beta) + \sum_{e\in E\backslash\{0\}} \sum_{\mu\in\mathbf{Z}^s} \gamma_{M\mu+e} \psi_e(M^r x - \beta - \mu)$$

Hence whenever $\beta \in \mathcal{B}$ so is $M\beta \in \mathcal{B}$.

Now, let α be any vector in $\mathbf{Z}^s\backslash\{0\}$. We wish to show $\alpha \in \mathcal{B}$. For this purpose we choose an integer m so that

$$\|M^{-m}\alpha\|_\infty < 1/2. \tag{2.15}$$

We express α in the form $\alpha = M\beta_1 + e$ for some $\beta_1 \in \mathbf{Z}^s$. If $e \in E\backslash\{0\}$ we conclude that $\alpha \in \mathcal{B}$, otherwise $e = 0$ and so $\alpha = M\beta_1$. Next, we write β_1 in the $\beta_1 = M\beta_2 + e_2$, $\beta_2 \in \mathbf{Z}^s$. Again, if $e_2 \in E\backslash\{0\}$ then $\beta_1 \in \mathcal{B}$ and so it follows from what we have shown above that $\alpha \in \mathcal{B}$. Otherwise, we obtain that $\alpha = M^2\beta_2$. We continue in this fashion until we conclude either $\alpha \in \mathcal{B}$

or $\alpha = M^m \beta_m$ for some $\beta_m \in \mathbf{Z}^s$. In the latter case (2.15) implies that $\beta_m = 0$ which is impossible since $\alpha \neq 0$. Thus α is indeed in $\mathcal{B}$ as claimed.

We finish the proof by showing that

$$\Gamma(M^r \bullet) = \psi(M^{r-1} \bullet) \in \overline{W} \tag{2.16}$$

for all $r \in \mathbf{Z}$. Since $M(W_e^k) = W_e^{k+1}$ it suffices to prove (2.16) for $r = 1$. According to Lemma 2.1

$$\psi = n^{-1}\bar{\rho}sc^{-1}\psi + \Omega \tag{2.17}$$

where $\Omega \in W^\circ$.

First we suppose $|\rho| < \sqrt{n}$. Then from (2.17), we have for any $\ell \in \mathbf{Z}_+$

$$sc^{-\ell}\psi = n^{-1}\bar{\rho}sc^{-\ell-1}\psi + sc^{-\ell}\Omega.$$

Thus we get

$$\psi = (n^{-1}\bar{\rho})^\ell sc^{-\ell}\psi + \Omega_\ell^+$$

where

$$\Omega_\ell^+ := \Omega + n^{-1}\bar{\rho}sc^{-1}\Omega + \cdots + (n^{-1}\bar{\rho})^{\ell-1}sc^{-\ell+1}\Omega.$$

Now, $\Omega_\ell^+ \in W$ and

$$\|\psi - \Omega_\ell^+\|_2^2 = (n^{-1}\bar{\rho})^{2\ell}\|sc^{-\ell}\psi\|_2^2 = (n^{-1}\bar{\rho}^2)^\ell\|\psi\|_2^2 \to 0, \quad \text{as} \quad \ell \to \infty.$$

Thus $\psi \in \overline{W}$.

When $|\rho| > \sqrt{n}$ we return to (2.17) and write it in the form

$$\psi = n\bar{\rho}^{-1}sc\psi - n\bar{\rho}^{-1}sc\Omega. \tag{2.18}$$

As above, by interating (2.18) we get for any $\ell \in \mathbf{Z}_+$

$$\psi = (n\bar{\rho}^{-1})^\ell sc^\ell\psi + \Omega_\ell^-$$

where

$$\Omega_\ell^- := -(n\bar{\rho}^{-1})(sc\Omega + \cdots + (n\bar{\rho}^{-1})^{\ell-1}sc^\ell\Omega) \in W.$$

Therefore

$$\|\psi - \Omega_\ell^-\|_2^2 = (n\bar{\rho}^{-1})^{2\ell}n^{-\ell}\|\psi\|_2^2 = (n\bar{\rho}^{-2})^\ell\|\psi\|_2^2 \to 0, \quad \text{as} \quad \ell \to \infty$$

since $|n\bar{\rho}^{-2}| < 1$ in this case. Hence in all cases we have $\psi \in \overline{W}$. This proves the corollary.

3*

As yet, we have not used one of the properties of the Laurent series $b(w)$ assumed in Theorem 2.1, namely,

$$\text{meas } \{w : b(w) = 0, \ w \in [-\pi, \pi]^s\} = 0. \tag{2.19}$$

In what follows it becomes essential. Note that when $l(w)$ satisfies (2.19), then $d(w)$ given by (2.9) also satisfies this condition.

By definition, we have $\Gamma = [d, \varphi]$. Thus, the last claim of Theorem 2.1, given in (c) will follow from Corollary 2.1 and the following general fact.

PROPOSITION 2.1. *Let $\varphi \in L^2(\mathbb{R}^s)$ satisfy*

$$\|[c, \varphi]\|_2 \le K \|c\|_2,$$

for some constant $K < \infty$ and all $c \in \ell^2(\mathbb{Z}^s)$, (so that $V^k(\varphi) \subseteq L^2(\mathbb{R}^s)$). If

$$\overline{\sum_{k \in \mathbb{Z}} V_k(\varphi)} = L^2(\mathbb{R}^s), \tag{2.20}$$

then

$$\overline{\sum_{k \in \mathbb{Z}} V_k(\Gamma)} = L^2(\mathbb{R}^s), \tag{2.21}$$

for any $\Gamma = [d, \varphi]$ with $d \in \ell^1(\mathbb{Z}^s)$ for which $d(w)$ satisfies (2.19).

Proof. It was observed in [5], see also [4] that for any $f, g \in L^2(\mathbb{R}^s)$,

$$\int_{\mathbb{R}^s} \bar{f}(x) g(x - \alpha) dx = 0, \tag{2.22}$$

all $\alpha \in \mathbb{Z}^s$, if and only if

$$\sum_{\beta \in \mathbb{Z}^s} \hat{f}(w + 2\pi\beta) \hat{g}(w + 2\pi\beta) = 0 \tag{2.23}$$

a.e. $w \in [-\pi, \pi]^s$. Now, to show (2.21) we will show that any $f \in L^2(\mathbb{R}^s)$ for which

$$\int_{\mathbb{R}^s} f(x) \Gamma(M^k x - \alpha) dx = 0, \quad k \in \mathbb{Z}, \ \alpha \in \mathbb{Z}^s, \tag{2.24}$$

is identically zero. According to the equivalence of (2.22) and (2.23) (applied to the functions $sc^{-k}f$ and Γ) we see that (2.24) implies that for all $k \in \mathbb{Z}$

$$\sum_{\beta \in \mathbb{Z}^s} (sc^{-k}\hat{f})(w + 2\pi\beta) \hat{\Gamma}(w + 2\pi\beta) = 0, \tag{2.25}$$

a.e., $w \in [-\pi, \pi]^s$. Since $\hat{\Gamma}(w) = d(w)\hat{\varphi}(w)$, and $d(w) \neq 0$, a.e., $w \in [-\pi, \pi]^s$ we conclude from (2.25) that

$$\sum_{\beta \in \mathbf{Z}^s} (sc^{-k}\hat{f})(w + 2\pi\beta)\hat{\varphi}(w + 2\pi\beta) = 0, \tag{2.26}$$

a.e., $w \in [-\pi, \pi]^s$. But then we can invoke the equivalence fo (2.22) and (2.23) one more time to obtain

$$\int_{\mathbf{R}^s} f(x)\varphi(M^k x - \alpha)dx = 0, \quad k \in \mathbf{Z}, \quad \alpha \in \mathbf{Z}^s.$$

But, according to (2.20), this implies $f = 0$ which proves the theorem.

3. Examples

In this section we describe some specific examples of Theorem 2.1. Specifically, we show that the cases studied in [2] and [4] are subsumed by this result.

Example 3.1 (Cube Spline). Let X be an $s \times m$ integer matrix with columns $x^1, \ldots, x^m \in \mathbf{Z}^s \backslash \{0\}$ and rank $s \leq m$. The trigonometric polynomial

$$a(w) = 2^{-m+s} \Pi_{j=1}^m (e^{iw \cdot x^j} + 1)$$

satisfies (2.2) with $M = \mathrm{diag}\,(2, 2, \ldots, 2), b(w) = \Pi_{j=1}^m (e^{-iw \cdot x^j} - 1)$, and $\rho = 2^{-m+s}$. Since $n = \det M = 2^s$ we have $\rho^2 = 2^{-2m+2s} < 2^s = n$.

The corresponding refinable function φ is the cube spline C_X whose Fourier transform is defined to be

$$\hat{C}_X(w) := \Pi_{j=1}^m \left(\frac{e^{iw \cdot x^j} - 1}{iw \cdot x^j} \right)$$

Equivalently, C_X has an alternative distributional definition

$$\int_{\mathbf{R}^s} C_X(w) f(w) dw := \int_{[0,1]^m} f(Xt) dt.$$

Obviously, C_X is of compact support and it is known that it admits multiresolution when X is unimodular, [2]. Recall that a matrix is unimodular whenever any nonsingular $s \times s$ submatrix has determinant ± 1.

Example 3.2 (Polyharmonic B-Spline).

The polyharmonic B-spline φ, normalized to have integral one has a Fourier transform given by

$$\hat{\varphi}(w) = 2^{-2r} \left(\sum_{j=1}^s \sin^2 w_j/2 \right)^r / \|w\|_2^{2r}$$

where we assume $r > s/2$.

To insure that φ satisfies (2.1) for the matrix $M = \mathrm{diag}\,(2, 2, \ldots, 2)$ we choose

$$a(w) = 2^{s-2r} \left(\sum_{j=1}^{s} \sin^2 w_j \right)^r \Big/ \left(\sum_{j=1}^{s} \sin^2 w_j/2 \right)^r, \quad w = (w_1, \ldots, w_s) \in \mathbb{R}^s.$$

Finally to guarantee that (2.2) holds we choose

$$b(w) = \left(\sum_{j=1}^{s} \sin^2 w_j/2 \right)^r,$$

and $\rho = 2^{s-2r}$. In this case $n = \det M = 2^s$ and so $\rho < \sqrt{n}$ because $s < 2r$, by hypothesis. Hence Theorem 2.1 applies in this case too.

References

[1] G. Battle, *A block spin construction of ondelettes: Part I: LeMarie Functions*, Commun. Math. Phys. **110** (1987), pp. 601–615.

[2] C.A. Micchelli, *Using the refinement equation for the construction of pre-wavelets*, Numerical Algorithms 1 (1991), pp. 75–116.

[3] C.A. Micchelli, *Banded matrices with banded inverses* Journal of Computational and Applied Mathematics, (to appear).

[4] C.A. Micchelli, C. Rabut, and F.I. Utreras, *Using the refinement equation for the construction of pre-wavelets III: elliptic splines* Numerical Algorithms, (to appear).

[5] R.Q. Jia and C.A. Micchelli, *Using the refinement equations for the construction of pre-wavelets II: Powers of Two Curves and Surfaces*, P.J. Laurent, A. Le Méhauté and L.L. Schumaker (eds), Academic Press, New York, 1991., (to appear).

[6] W. Rudin, *Functional Analysis*, McGraw Hill Book Company, New York, 1973.

T.J. Watson Research Center
P.O. Box 218
Yorktown Heights, NY 10598
USA

STRONG ASYMPTOTICS
FOR ORTHOGONAL POLYNOMIALS

E.A.Rakhmanov

Introduction

The purpose of this paper is to give a self contained and short proof of Bernstain-Szegö type asymptotics for orthogonal polynomials

$$H_n(x) = a_n x^n = \cdots : \int_{\mathbb{R}} H_n(x) H_m(x) h(x) dx = \delta_{n,m},$$

associated with weight function $h(x) = e^{-|x|^\rho}$, where $\rho > 1$. That is

THEOREM 1. *We have as* $n \to \infty$

$$a_n = (2\pi)^{-1/2} \exp\{-\frac{2n+1}{2\rho} \log \frac{2n+1}{e\alpha 2^{\rho+1}} + O(n^{-1/3})\}, \tag{i}$$

$$\int_{-x_n}^{x_n} | (\pi/2)^{1/2} h(x)^{1/2} H_n(x) - (x_n^2 - x^2)^{-1/4} \cos \Phi_n(x) |^2 \, dx = O(n^{-1/3}), \tag{ii}$$

$$(\pi/2)^{1/2} h(x)^{1/2} (x_n^2 - x^2)^{1/4} H_n(x) - \cos \Phi_n(x) \to 0, \tag{iii}$$

uniformly with respect to x, *where* $| x/x_n | \le 1 - \epsilon$ *with an arbitrary* $\epsilon > 0$. *Here*

$$\alpha = \alpha_\rho = \pi^{-1/2} \Gamma(\frac{\rho+1}{2}) / \Gamma(\frac{\rho}{2}), \qquad x_n = (\frac{n+1/2}{\alpha})^{1/\rho} \tag{1.0}$$

$$\Phi_n(x) = (n+1/2) \int_x^{x_n} (1 - x^\rho/t^\rho)(x_n^2 - t^2)^{-1/2} dt - \pi/4 \tag{2.0}$$

$\Phi_n(-x) = \pi n - \Phi_n(x)$, $x \in [0, x_n]$; Γ *is the Euler gamma-function.*

In this introduction we make a few short remarks on the problem. A general approach to the problem of asymptotics for the orthogonal polynomials associated with a weight function $h(x) = e^{-2\varphi(x)}$ on the real line was developed in the papers of author [9]–[11]. The two main its components were connected with introducing and investigation of the two following objects

    (A)   A family of equilibrium measures $\lambda_{\tau,\varphi}$, $\tau \in \mathbb{R}_+$ corresponding to the function (external field) φ and

    (B)   A sequence of polynomials $T_n(x)$ whose zeros are distributed uniformly with respect to $\lambda_{n+1/2,\varphi}$.

The polynomials T_n are proved to be close to the orthonormal polynomials H_n related to weight h and this property is the key point of the
whole approach. In [9]–[11] we studied logarithmic asymptotics for orthogonal polynomials and this required a rough bounds for T_n only. We note that
V.S.Bujarov has extended those results to a more general class of weights in
[1].

The general concept of the equilibrium measure in the external field on
$\mathbb{R}$ was introduced by Gonchar and Rakhmanov in [2] in connection with the
problem of contracted zero distribution (or equivalently the contracted n-th
root asymptotic; this is a weaker form of asymptotic relation). At the same
time N.H. Mhascar and E.B.Saff [7] working independently considered also
the problem of contracted zero distribution (for exponential weights) and
developed for this problem an approach similar to stated above.

For the study of strong asymptotics for orthogonal polynomials we need
some precise asymptotic formulae for polynomials T_n. The most part of this
paper is actually devoted to investigation of these polynomials.

Theorem 1 was announced at Segovia conference on Orthogonal polynomials and their applications (1986) and in the proceedings of the conference
[3] together with a general discussion of the problem as a whole and a few
details about the two main points of the proof (that is (A) and (B)). The
first version of the proof was not published because of his size.

At the same time D.S.Lubinsky and E.B.Saff [4] have proved assertions
(i) and (ii) of Theorem 1 for a somewhat general class of weights but without
estimate of error. Their approach includes a consideration of (A) based on
[10] and [8] and

(C) *Relative approximation of the weight by polynomials.*

instead of (B) above. In [5] D.S.Lubisky has developed this approach and
has also proved uniform asymptotics (iii) for a class of weights including
$h(x) = \exp(-|x|^\rho)$ with $\rho > 3$.

Finally we note that for the case $h(x) = \exp(-|x|^\rho)$ with $\rho = 2, 4, 6, \ldots$
a classical method is available for the study of asymptotics of orthogonal
polynomials based on differential equations for them. Developing this approach P.Nevai investigated the case $\rho = 4$ in [5] before all the results stated
above; R.S.Sheen extended this to $\rho = 6$. P.Nevai formulated also the correct
conjecture for all ρ.

1. The family of equilibrium measures $\lambda_\tau, \tau \in \mathbb{R}_+$ associated with the function (external field) $\varphi(x) = |x|^\rho /2$

Here we use a formal way of introducing λ_τ based on some explicit formulae.

We will need some notations for constants and functions associated with

$\varphi(x) = |x|^\rho /2$. In what follows they will be used without further explanations. We set

$$\beta_\tau = (\tau/\alpha)^{1/\rho}, \text{ where } \alpha = \pi^{-1/2}\Gamma(\tfrac{\rho+1}{2})/\Gamma(\tfrac{\rho}{2}), \tag{1.1}$$

$$\nu(x) = \alpha x^\rho, x \geq 0 \text{ is the inverse function to } \beta_x, x \geq 0 \tag{1.2}$$

$$\Delta_\tau = [-\beta_\tau, \beta_\tau] \text{ is the segment of } \mathbb{R} \text{ for any } \tau > 0$$

Let γ_τ be the Robin distribution for Δ_τ and $g_\tau(x)$ be the Green function for the exterior of Δ_τ with pole at infinity. We have $g_\tau = 0, x \in \Delta_\tau$, and

$$d\gamma_\tau(x) = \pi^{-1}(\beta_\tau^2 - x^2)^{-1/2}dx, \qquad x \in \Delta_\tau \tag{1.3}$$

$$g_\tau(x) = \int_{\beta_\tau}^{|x|} (t^2 - \beta_\tau^2)^{-1/2}dt, \qquad x \in \mathbb{R}\backslash\Delta_\tau \tag{1.4}$$

$$g_\tau(z) = \log(2/\beta_\tau) - V^{\gamma_\tau}(z) \tag{1.5}$$

where $V^\mu(z) = \int \log |z - t|^{-1}\, d\mu$ is the logarithmic potential of measure μ; sometimes we denote a measure μ by $d\mu$. Further we set

$$w_\tau = \int_0^\tau \log(2/\beta_t)dt = -\frac{\tau}{\rho}\log\frac{\tau}{e\alpha 2^\rho} \tag{1.6}$$

$$\lambda'_\tau(x) = \pi^{-1}\int_{\nu(x)}^\tau \frac{dt}{(\beta_t^2 - x^2)^{1/2}} = \pi^{-1}\int_x^{\beta_\tau} \frac{\nu'(t)dt}{(t^2 - x^2)^{1/2}} \tag{1.7}$$

$$d\lambda_\tau(x) = \lambda'_\tau(x)dx, \qquad x \in \Delta_\tau$$

Hence λ_τ is a positive and absolutely continuous measure on Δ_τ. Comparing (1.7) and (1.3) we see that this measure can be represented as follows

$$\lambda_\tau = \int_0^\tau \gamma_t dt \tag{1.8}$$

This means that we have $\int f d\lambda_\tau = \int_0^\tau (\int f d\gamma_t)dt$ for any function f on Δ_τ if these integrals exist. Thus we obtain the relations

$$V^{\lambda_\tau}(z) = \int_0^\tau V^{\gamma_t}(z)dt; \qquad |\lambda_\tau| = \tau \tag{1.9}$$

LEMMA 1. The following relations are true:

$$\varphi(x) = \int_0^\infty g_t(x)dt, \qquad x \in \mathbb{R} \tag{i}$$

$$V^{\lambda_t}(x) + \varphi(x) = w_t, \qquad x \in \Delta_t \tag{ii}$$

$$V^{\lambda_\tau}(x) + \varphi(x) - w_\tau = \int_\tau^{\nu(|x|)} g_t(x)dt, \qquad x \in \mathbb{R}\backslash\Delta_\tau \qquad \text{(iii)}$$

$$V^{\lambda_\tau}(x) + \varphi(x) - w_\tau \geq C\tau\beta_\tau^{-1} \mid x \mid^{-1/2} (\mid x \mid -\beta_\tau)^{3/2}, \qquad x \in \mathbb{R}\backslash\Delta_\tau \quad \text{(iv)}$$

Proof. Both functions in (i) are even, so it is enough to prove it for $x \geq 0$. According to (1.4) we have $g_t'(x) = 0$, $t \geq \nu(x)$ and $g_t'(x) = (x^2 - \beta_t^2)^{-1/2}$, $t < \nu(x)$. Taking this into account we obtain

$$\frac{d}{dx}\int_0^\infty g_t(x)dt = \int_0^{\nu(x)}(x^2 - \beta_t^2)^{-1/2}dt = \int_0^x \nu'(t)(x^2 - t^2)^{-1/2}dt =$$

$$= \int_0^1 (1-t^2)^{-1/2}\nu'(xt)dt = \rho\alpha_\rho x^{\rho-1}\int_0^1 (1-t^2)^{-1/2}t^{\rho-1}dt = \frac{\rho}{2}x^{\rho-1} = \varphi'(x).$$

On the other hand we have $\varphi(0) = \int_0^\infty g_t(0)dt = 0$, and (i) follows.

In turn, it follows from (i) and (1.5) that

$$\varphi(x) = \int_0^\tau g_t(x)dt + \int_\tau^\infty g_t(x)dt = \int_0^\tau (\log\frac{2}{\beta_t} - V^{\gamma_t}(x))dt + \int_\tau^\infty g_t(x)dt$$

Now we use (1.6), (1.9) and obtain for any $x \in \mathbb{R}, \tau > 0$

$$V^{\lambda_\tau}(x) + \varphi(x) - w_\tau = \int_\tau^\infty g_t(x)dt = \int_\tau^{\nu(|x|)} g_t(x)dt$$

If $x \in \Delta_\tau$, the right hand side vanishes and (ii) follows. In the case $x \in \mathbb{R}\backslash\Delta_\tau$ we have $g_t(x) = 0$ for $t > \nu(|x|)$ and (iii) is also proved. Finally, let $x > \beta_\tau$, then we have from (1.4)

$$g_\tau(x) = \int_{\beta_\tau}^x \frac{dt}{(t^2 - x^2)^{1/2}} > (2x)^{-1/2}\int_{\beta_\tau}^x \frac{dt}{(t - x)^{1/2}} = (2(x - \beta_\tau)/x)^{1/2}$$

Hence $\int_\tau^{\nu(x)} g_t(x)dt \geq (2/x)^{1/2}\int_\tau^{\nu(x)}(x - \beta_t)^{-1/2}dt = (2/x)^{1/2}\int_{\beta_\tau}^x \frac{\nu'(t)dt}{(t-x)^{1/2}}$. Using inequality $\nu'(x) \geq \nu'(\beta_t)$ for $x \geq \beta_\tau$, we deduce (iv) with $C = (8/9)^{1/2}\rho\alpha_\rho$. The proof is completed.

LEMMA 2. *The function* $\lambda_\tau'(x)$ *defined by (1.7) can also be represented as*

$$\lambda_\tau'(x) = \pi^{-1}\tau\rho\int_{|x|}^{\beta_\tau}(\mid x \mid^{\rho-1}/t^\rho(\beta_\tau^2 - t^2)^{-1/2}dt. \qquad (1.9)$$

Proof. Let us define $a(x)$ to be equal the right hand side of (1.9). Now we have to prove that $\lambda_\tau'(x) = a(x)$, $x \in (0,\beta)$. We show that each of these two functions satisfies the following differential equation:

$$xf'(x) = (\rho - 1)f(x) - \pi^{-1}\rho\tau(\beta_\tau^2 - x^2)^{-1/2}, \quad x \in (0,\beta). \qquad (1.10)$$

with respect to a unknown function $f(x)$. For $f = a$ this can easily be verified by definition. For $f = \lambda'_\tau$ we will use relation

$$\pi(\alpha\rho)^{-1}\lambda'_\tau(x) = \int_x^{\beta_\tau} t^{\rho-1}(t^2 - x^2)^{-1/2}dt \tag{1.11}$$

$$\pi(\alpha\rho)^{-1}\lambda\tau'(x) = \beta_\tau^{\rho-2}(\beta_\tau^2 - x^2)^{1/2} - (\rho - 2)\int_x^{\beta_\tau} t^{\rho-3}(t^2 - x^2)^{1/2}dt \tag{1.12}$$

$$\pi(\alpha\rho)^{-1}(\lambda'_\tau)'(x) =$$
$$= -x\beta_\tau^{\rho-2}(\beta_\tau^2 - x^2)^{-1/2} + x(\rho - 2)\int_x^{\beta_\tau} t^{\rho-3}(t^2 - x^2)^{-1/2}dt \tag{1.13}$$

(1.11) follows by (1.7), then this implies (1.12) by integrating by parts and this implies (1.13) by taking the derivative in x.

We multiply (1.11) by $\rho - 2$ then multiply (1.13) by x and take the difference. After straightforward transformations (using again (1.12)) this yields (1.10) for $f = \lambda'_\tau$.

It remains to note that $a(x) \to 0$ as $x \to \beta - 0$ and the same is true for λ'_τ. Equation (10) has a unique solution with this property, and the proof is completed.

We remark that by (1.1) we obtain a new expression for x_n from Theorem 1 (see (1.0))

$$x_n = \beta_{n+1/2} \tag{1.14}$$

As a corollary of Lemma 2 we also have another representation for $\Phi_n(x)$ defined by (2.0):

$$\Phi_n(x) = \pi \int_x^{\beta_{n+1/2}} d\lambda_{n+1/2} - \pi/4. \tag{1.15}$$

To prove this one can substitute $d\lambda_{n+1/2}$ with its expression from Lemma 2 and then integrate by parts.

Finally we denote

$$v(x) = \pi^{-1}\rho\int_{|x|}^1 (|x|^{\rho-1}/t^\rho(1 - t^2)^{-1/2}dt, \quad x \in [-1, 1] \tag{1.16}$$

Making use of substitution $t = \beta_\tau\xi$, $\xi \in [-1, 1]$ in the integral on the right hand side of (1.9) and then putting $y = \beta_\tau x$ there, we obtain

$$\lambda'_\tau(x) = \beta_\tau^{-1}\tau v(\beta_\tau^{-1}x), \quad x \in \Delta_\tau. \tag{1.17}$$

2. Outline of the proof

2.1 Orthogonal polynomials Q_n associated with truncated weight

Let $Q_n(x, \tau) = b_n x^n + \cdots$ be a sequence of orthogonal polynomials associated with the weight $h(x) = \exp(-|x|^p)$ on the finite segment Δ_τ:

$$\int_{\beta_\tau}^{-\beta_\tau} Q_n(x,\tau) Q_m(x,\tau) h(x) dx = \delta_{n,m}$$

An important intermediate result in our approach is the strong asymptotics for the sequence of polynomials $Q_n(x, \tau_n)$, where the segment of orthogonality $\Delta_{\tau_n} = [-\beta_{\tau_n}, \beta_{\tau_n}]$ increases together with the degree of polynomials in such a way that

$$0 \leq n + 1/2 - \tau_n \leq M, n \in \mathbb{N} \tag{2.1}$$

(Theorem 2 below). After that we will show that polynomials $H_n(x)$ and $Q_n(x, \tau)$ are close to each other in the strong sense (in particular, uniformly on the n-th segment of orthogonality). So we conclude that for H_n the same asymptotics are valid.

It is convenient now to transform statements (i)–(iii) of Theorem 1 to an equivalent form including some new constants and functions.

Suppose that $n \in \mathbb{R}, \tau \geq n + 1/2$ are fixed. We denote

$$\mu_{n,\tau} = \lambda_\tau + (n - \tau + 1/2)\gamma_\tau \tag{2.2}$$

$$m_{n,\tau} = w_\tau + (n - \tau + 1/2)\log(2/\beta_\tau) \tag{2.3}$$

$$c_n(\tau) = (2\pi)^{-1/2} \exp(m_{n,\tau}) \tag{2.4}$$

$$\phi_n(x,\tau) = \pi \int_x^{\beta_\tau} d\mu_{n,\tau}(t) - \pi/4 \tag{2.5}$$

$$A(x,\tau) = (2\pi)^{-1/2}(\beta_\tau^2 - x^2)^{-1/4} h(x)^{-1/2} \tag{2.6}$$

$$R_n(x,\tau) = 2A(x,\tau)\cos\phi_n(x,\tau) \tag{2.7}$$

We denote by $L_2(h, \tau)$ the Hilbert space of all complex-valued functions f on Δ_τ with the finite norm $\|f\|$ generated by the scalar product

$$(f,g)_\tau = \int_{\Delta_\tau} f(x)\overline{g(x)} h(x) dx; \qquad \| f \|_\tau = (f, f)_\tau. \tag{2.8}$$

We note that the case $\tau = \infty$ is also included, so that $\Delta_\infty = \mathbf{R}$. The orthogonality condition for H_n and Q_n may now be written as

$$(Q_n, Q_m)_\tau = \delta_{n,m}, \qquad (H_n, H_m)_\infty = \delta_{n,m}.$$

By $\mathbb{P}_n$ we denote the set of polynomials whose degree does not exceed n. Finally we denote

$$d_n(\tau) = \min_{Q \in \mathbb{P}_n} \| Q - R_n \|_\tau^2 \tag{2.9}$$

$$e_n(\tau) = \pi^{-1} \int_{\Delta_\tau} \cos(2\phi_n(x, \tau))(\beta_n^2 - t^2)^{-1/2} dt \tag{2.10}$$

$$\hat{Q}_n(x, \tau) = c_n(\tau)^{-1} b_n(\tau) Q_n(x) \tag{2.11}$$

In this notation we obtain a new reduction of Theorem 1.

THEOREM 1. *Suppose that* $\tau_n = n + 1/2$. *Then we have as* $n \to \infty$

$$a_n/c_n(\tau_n) = 1 + O(n^{-1/3}) \tag{i}$$

$$\| H_n(x) - R_n(x, \tau_n) \|_{\tau_n}^2 = O(n^{-1/3}), \tag{ii}$$

$$\beta_{\tau_n}^{1/2} h(x)^{1/2} | H_n(x) - R_n(x, \tau_n) | \to 0 \tag{iii}$$

uniformly with respect to x, *where* $| x/\beta_{n+1/2} | \leq 1 - \epsilon, \epsilon > 0$.

It follows by (1.14) and (1.15) that (i)–(iii) here are equivalent to those in the first form of the Theorem 1 given in the introduction. To prove this theorem we first prove a similar theorem for the polynomials associated with truncated weight.

THEOREM 2. *Suppose that condition (2.1) is satisfied for the sequence* τ_n. *Then we have as* $n \to \infty$

$$b_n(\tau_n)/c_n(\tau_n) = 1 + O(n^{-1/3}), \tag{i}$$

$$\| Q_n(x, \tau_n) - R_n(x, \tau_n) \|_{\tau_n}^2 = O(n^{-1/3}), \tag{ii}$$

$$\beta_{\tau_n}^{1/2} h(x)^{1/2} | Q_n(x, \tau_n) - R_n(x, \tau_n) | \to 0 \tag{iii}$$

uniformly with respect to x, *where* $| x/\beta_{\tau_n} | \leq 1 - \epsilon, \epsilon > 0$.

The first step in the proof of Theorem 2 is the following lemma which contains some formal properties of $R_n(x, \tau)$ and of the constant $c_n(\tau)$.

LEMMA 3. *For any* $n \in \mathbf{N}, \tau > 0$ *we have*

$$d_n(\tau) = \parallel \hat{Q}_n - R_n \parallel_\tau^2 \tag{i}$$

$$c_n(\tau)^{-1} b_n(\tau) = 1 + e_n(\tau) - d_n(\tau) \tag{ii}$$

It follows from (i) that $\hat{Q}_n$ is the orthogonal projection of R_n in $L_2(h,\tau)$ to the subspace $\mathbf{P}_n$. Hence we have the following equivalent expression for (i)

$$\hat{Q}_n(x,\tau) = \int_{\Delta_\tau} R_n(t,\tau) K_n(t,x;\tau) h(t) dt \tag{i'}$$

where K_n is the reproducing Szegö kernel function (see §5).

The proof of Lemma 3 is presented in the next section 3. Now we continue the review of the components of the proof.

2.2 A construction of polynomials close to R_n

It follows from Lemma 3 that the investigation of asymptotics for truncated orthogonal polynomials Q_n can be reduced to the construction of some polynomials $T_n \in \mathbf{P}_n$, which are close to R_n.

The key point of our approach is the following construction of polynomials. For any $\tau \leq n + 1/2$ we denote

$$T_n(x,\tau) = c_n(\tau) \prod_{k=1}^{n} (x - x_k), \quad \text{where} \quad R_n(x_k,\tau) = 0, \tag{2.12}$$

Under condition $\tau \leq n + 1/2$ the function $R_n(x,\tau)$ has exactly n zeros $x_k = x_{k,n}(\tau), k = 1,2,\ldots,n$ inside Δ_τ because the zeros of $R_n(x,\tau)$ coincide with the zeros of $\cos\phi_n(x,\tau)$ and $\phi_n(x,\tau)$ decrease on Δ_τ from $\pi n + \pi/4$ to $-\pi/4$; see definitions (2.2) and (2.5). It follows also from the definitions that points x_k can be defined equivalently by the system of equations

$$\int_{x_k}^{x_{k+1}} d\mu_{n,\tau} = 1, \quad k = 1,\ldots,n-1; \quad \int_{-\beta_\tau}^{x_1} d\mu_{n,\tau} = \int_{x_n}^{\beta_\tau} d\mu_{n,\tau} = 3/4. \tag{2.13}$$

It is essential for (2.13) to have a solution satisfying $\mid \mu_{n,\tau} \mid = n + 1/2$, where $n \in \mathbf{N}$. We see that T_n really depend only on a measure $\mu_{n,\tau}$ and can be defined as polynomials whose zeros are distributed uniformly with respect to this measure.

LEMMA 4. *Let* τ_n *be a sequence satisfying (2.1), then we have*

$$h(x)^{-1/2} (\beta_{\tau_n}^2 - x^2)^{1/4} \mid R_n(x,\tau_n) - T_n(x) \mid \leq C\delta_n(y) \tag{i}$$

where $x \in \Delta_{\tau_n}$, $y = x/\beta_{\tau_n} \in [0,1]$ and

$$\delta_n(y) = \frac{1}{n^{\rho-1}(1+n \mid y \mid)} + \frac{1}{1+n^{2/3}(1- \mid y \mid)}$$

$$\int_{\mathbb{R}\backslash \Delta_{\tau_n}} T_n^2(x)h(x)dx = O(n^{-1/3}) \qquad \text{(ii)}$$

$$\int_{\mathbb{R}} T_n^2(x)h(x)dx = 1 + O(n^{-1/3}) \qquad \text{(iii)}$$

$$e_n(\tau_n) = O(n^{-1/3}) \qquad \text{(iv)}$$

The proof of Lemma 4 is contained in section 4. In section 4 we obtain an integral representation for the function $\eta = \log(T_n/R_n)$ on Δ_τ, which may have an independent interest.

Assertions (i), (ii) of both Theorems 1 and 2 are immediate corollaries of Lemmas 3, 4

2.3. Proof of L_2 - asymptotics for Q_n and H_n

Using Lemma 3, (i) and Lemma 4, (i) we obtain

$$d_n(\tau_n) = \| \hat{Q}_n(x,\tau_n) - R_n(x,\tau_n) \|_{\tau_n}^2 \le \| T_n(x) - R_n(x,\tau_n) \|_{\tau_n}^2 \le$$
$$\le C \int_{-1}^1 \delta_n^2(y)(1-y^2)^{-1/2}dy = O(n^{-1/3}). \qquad (2.15)$$

Using this and Lemma 3, (ii), Lemma 4, (iv) all together we prove (i) of Theorem 2. In turn by Theorem 2,(i) and the estimate $\| \hat{Q}_n - R_n \|_{\tau_n}^2 = O(n^{-1/3})$ from (2.15), we obtain (ii) of Theorem 2.

Now, the extremal property of the leading coefficients of orthonormal polynomials (applied to $b_n(\tau_n)$) implies

$$a_n^2/b_n^2(\tau_n) \le \| H_n \|_{\tau_n}^2 < 1. \qquad (2.16)$$

Hence, by Theorem 2,(i) we have $a_n^2/c_n^2(\tau_n) \le 1 + O(n^{-1/3})$. On the other hand by Lemma 4, (iii) and the extremal property of a_n we have the inverse inequality $a_n^2/c_n^2(\tau_n) \ge 1 + O(n^{-1/3})$. These two inequalities taken together yield (i) of Theorem 1. In turn by Theorem 1, (i) we have (using a usual trick) $\| T_n - H_n \|_\infty^2 = \| T_n \|_\infty^2 - 2(T_n, H_n) + \| H_n \|_\infty^2 = 1 + O(n^{-1/3}) - 2c_n(\tau_n)/a_n + 1 = O(n^{-1/3})$. Hence $\| T_n - H_n \|_{\tau_n}^2 = O(n^{-1/3})$; by (2.15) the same is true for $T_n - R_n$ and (ii) of Theorem 1 follows.

Let us remark the following three relations obtained during the above proof

$$a_n^2/b_n^2(\tau_n) = 1 + O(n^{-1/3}) \qquad (2.17)$$

$$\| T_n - H_n \|_\infty^2 = O(n^{-1/3}) \tag{2.18}$$

$$\int_{\mathbb{R}\backslash\Delta_{n+1/2}} H_n^2(x)h(x)dx = O(n^{-1/3}) \tag{2.19}$$

((2.19) follows by (2.17) and (2.18) because the integral in the left hand side of (2.19) is equal to $1- \| H_n \|_{T_n}^2$).

2.4. Uniform asymptotics

The final part of the proof of the theorems establishing uniform asymptotics includes something hardly unexpected. We use Lemma 3, (i') to treat polynomials Q_n and then Koraus' identity to extend the result to H_n. Related calculations are presented in section 6. It is essential there that the precise estimates in L_2 – asymptotics have already been obtained. As an auxiliary result we need in sec.6 some simple bounds for Szegö' kernel function K_n. It is obtained before in sec.5. We shall be brief in details there.

3. The proof of Lemma 3

We need the following classical result on the boundary values of Cauchy type integrals. Let $\Delta = [\alpha,\beta]$, $f \in L_1(\Delta)$ and

$$F(z) = \int_\Delta (t-z)^{-1} f(t)dt, \ z \in \mathbb{C}\backslash\Delta.$$

Then $F(z)$ has boundary values $F^+(x)$ and $F^-(x)$ from upper and lower halfplanes respectively almost everywhere on Δ and

$$F^\pm(x) = \int_\Delta (t-z)^{-1} f(t)dt \pm \pi \ i \ f(x) \tag{3.1}$$

where the integral is taken in the principal value (Plemelj-Sokhotsky formula). Furthermore if f is continuous on (α,β) then $\Im F(z)$ admits a continuous extension to the whole boundary of the domain $\mathbb{C}\backslash\Delta$ with possible exception of the end points α,β and (3.1) for imaginary parts is valid for every $x \in (\alpha,\beta)$.

We introduce also the notation for the complex potential of a measure μ on $\Delta = [\alpha,\beta]$: $U^\mu(z) = V^\mu(z) + i\hat{V}^\mu(z) = \int \log(z-t)^{-1}d\mu(t)$, where $\hat{V}^\mu(z)$ is the conjugate harmonic function to $V^\mu(z)$ in $\mathbb{C}\backslash\Delta$. The function $U^\mu(z)$ is multivalued in $\mathbb{C}\backslash\Delta$ (as well as $\hat{V}^\mu(z)$); we will consider its single valued branch in the domain $\mathbb{C}\backslash(-\infty,\beta]$ normalized by the condition $\hat{V}^\mu(x) = 0, x \in (\beta,+\infty)$.

LEMMA 3.1. *Let $d\mu(x) = f(x)dx$ be an absolutely continuous measure on $\Delta = [\alpha, \beta]$ and its derivative f is continuous on (α, β). Then its complex potential has a continuous continuation to the boundary of $\mathbb{C}\backslash\Delta$ and*

$$(U^\mu)^\pm(x) = V^\mu(x) \mp \pi i \int_x^\beta d\mu(t)$$

Proof. It is sufficient to consider the imaginary parts of the left and right hand sides. Making use of the abbreviation $U^\mu = U = V + i\hat{V}$ we have $U'(z) = \int_\Delta (t-z)^{-1} f(t)dt$. Hence using (3.1) and the related remark we obtain $\frac{d}{dx}\hat{V}^\pm(x) = (\Im U')^\pm(x) = \pm\pi i f(x)$, $x \in (\alpha, \beta)$. Both functions $\hat{V}^\pm(x)$ are continuous on $[\alpha, \beta]$ and $\hat{V}^\pm(\beta) = 0$ due to our choice of the branch of $\hat{V}(z)$. The assertion of lemma follows. (Lemma holds without any restrictions on μ but in what follows we need just what we have proved).

Now we complete our notation (2.2)-(2.11) by

$$W_n(z, \tau) = c_n(\tau)(z^2 - \beta_\tau^2)^{-1/4} \exp(-U^{\mu_{n,\tau}}(z)), \quad z \in \mathbb{C}\backslash\Delta_\tau \qquad (3.2)$$

where the branch of $(z - \beta_\tau)^{-1/4}$ is supposed to be positive on $[\beta, +\infty)$ and U^μ is defined above. Taking also into account that $\mu_{n,\tau}(\Delta_\tau) = n + 1/2$ we see that W_n has a single valued analytic continuation to the whole domain $\mathbb{C}\backslash\Delta_\tau$ with the pole at infinity and the Laurent expansion of the form

$$W_n(z, \tau) = c_n(\tau)z^n + \dots \qquad (3.3)$$

This function has no zeros in $\mathbb{C}\backslash\Delta_\tau$.

LEMMA 3.2. *We have for the boundary values of W_n for $x \in \Delta_\tau$*

$$W_n^\pm(x, \tau) = A(x, \tau) \exp(\pm i\phi_n(x, \tau)), \qquad (i)$$

$$\Re W_n^\pm(x, \tau) = R_n(x, \tau)/2, \qquad (ii)$$

$$|W_n^\pm(x, \tau)|^{-2} = 2\pi(\beta_\tau^2 - x^2)^{1/2} h(x). \qquad (iii)$$

Proof. We use Lemma 1, (ii), (1.5) together with (2.2) and (2.3). They imply that $m_{n,\tau} - V^{\mu_{n,\tau}}(x) = \varphi(x)$, $x \in \Delta_\tau$. We also have

$$((z^2 - \beta_\tau^2)^{-1/4})^\pm(x) = (\beta_\tau^2 - x^2)^{-1/4} \exp(\mp\pi i/4), \quad x \in \Delta_\tau.$$

Using that and then (2.2)–(2.4), (2.6), and Lemma 3.1 we obtain (i). Now (ii) and (iii) follow by (i) and (2.5)–(2.7).

LEMMA 3.3. *Suppose* $q(x) = kx^n + \ldots \in \mathbf{P}_n$, $R_n(x) = R_n(x, \tau)$. *Then*

$$(q, R_n)_\tau = k/c_n(\tau)$$

Proof. Without loss of generality we can suppose that q has real coefficients. The function $u(z) = \Re(q(z)/W_n(z))$ is harmonic in $\bar{\mathbf{C}} \backslash \Delta_\tau$. By (ii) and (iii) of Lemma 3.2 we have

$$u^\pm(x) = \Re(q(x)W^\pm(x))/\mid W_n^\pm(x)\mid^2 = \pi q(x)R_n(x)(\beta_\tau^2 - x^2)^{1/2}h(x)$$

In particular u has continuous continuation to $\bar{\mathbf{C}}$. Using appropriate version of Poisson formula $u(\infty) = \pi^{-1} \int_{\Delta_\tau} (\beta_\tau^2 - t^2)^{-1/2} u(t) dt$ we obtain

$$u(\infty) = \int_{\Delta_\tau} q(x) R_n(x) h(x) dx = (R_n, q)_\tau$$

On the other hand we have $u(\infty) = k/c_n(\tau_n)$ by (3.3) and Lemma follows.

Now we can complete the proof of Lemma 3. By the orthogonality conditions for Q_n and (2.11) we obtain that $(\hat{Q}_n, q)_\tau = k/c_n(\tau)$ for any polynomial $q(x) = kx^n + \ldots$. Together with Lemma 3.3 this yields the equality $(\hat{Q}_n - R_n, q)_\tau = 0$ for any $q \in \mathbf{P}_n$. This for any $q \in \mathbf{P}_n$ implies

$$\| q - R_n \|_\tau^2 = \| \hat{Q}_n - R_n \|_\tau^2 + \| q - \hat{Q}_n \|_\tau^2 .$$

This yields (i) of Lemma 3. Further for any $z \in \mathbf{C}$ we have $2(\Re z)^2 = \Re(z^2) + |z|^2$. By this and Lemma 3.2, (ii) we obtain $R_n^2(x) = 2 \mid W_n(x) \mid^2 + 2\Re W_n^2(x)$, $x \in \Delta_\tau$ and further taking into account (i) and (iii) of Lemma 3.2

$$\| R_n \|_\tau^2 = 1 + \pi^{-1} \int_{\Delta_\tau} (\cos 2\phi_n(x))(\beta_n^2 - x^2)^{-1/2} dx = 1 + e_n(\tau)$$

On the other hand by Lemma 3.3 we have

$$d_n(\tau) = \| \hat{Q}_n - R_n \|_\tau^2 = \| R_n \|_\tau^2 - 2(\hat{Q}_n, R_n)_\tau + \| \hat{Q}_n \|_\tau^2 = \| R_n \|_\tau^2 - b_n^2(\tau)/c_n^2(\tau)$$

Together with what was said above it implies (ii) of Lemma 3. The proof is completed.

4. Integral representations for the functions
$\eta = \log(T_n/R_n)$ and $\theta = \log(T_n/W_n)$ and proof of Lemma 4

For $n \in \mathbf{N}, \tau \leq n + 1/2$ we define (see (2.6), (2.14), (3.2))

$$\eta(x) = \log(T_n(x)/R_n(x)), \qquad x \in \Delta = [-\beta, \beta], \tag{4.1}$$

$$\theta(z) = \log(T_n(z)/W_n(z)), \qquad z \in \bar{\mathbb{C}} \backslash \Delta; \qquad \theta(\infty) = 0. \qquad (4.2)$$

where $\beta = \beta_\tau$. In the following we drop the dependence of τ. By continuity of $\lambda'(x)$ and definitions it follows that $\eta(x)$ is continuous on $(-\beta, \beta)$ and $\eta(x) \to -\infty$ as $x \to \pm\beta$. The function $\theta(z)$ is holomorphic (analytic and single valued) in $\bar{\mathbb{C}} \backslash \Delta$ by the properties of W. The normalization $\theta(\infty) = 0$ is correct by (2.12) and (3.3).

The first purpose of this section is to obtain integral representations for these two functions (Theorem 3 below). We note that the method developed can be directly applied to any measure $d\lambda = \lambda' dx$ on $\Delta = [\alpha, \beta]$ under the following condition. There exist $p \geq 0$ points $\alpha = x_0 < x_1 < \ldots < x_p < x_{p+1} = \beta$ such that $\lambda'(x)$ is positive and analytic in $(x_{k-1}, x_k), k = 1, \ldots, p$.

In our case by (1.7), (2.2) we have

$$\mu_n'(x) = \pi^{-1} \int_{|x|}^{\beta} \nu'(t)(t^2 - x^2)^{-1/2} dt + \pi^{-1}(n + 1/2 - \tau)(\beta^2 - x^2)^{-1/2}, \quad (4.3)$$

where $\nu(x) = \alpha_\rho x^p$, $x \in \Delta$. Hence we obtain two intervals

$$\Delta^- = (-\beta, 0) \qquad \text{and} \qquad \Delta^+ = (0, \beta),$$

where μ' is analytic and positive. The same is true for

$$\phi(x) = \pi \int_x^{\beta} d\mu(t) - \pi/4 \qquad x \in \Delta. \qquad (4.4)$$

We will consider the analytic continuation of these two functions from Δ^+ and Δ^- to some domains in the upper halfplane. It is convenient to introduce immediately some special family of those domains. For a fixed $p > 0$ we denote

$$\delta = p\tau^{-1} \log \tau, \qquad (4.5)$$

$$b = \beta(1 - \delta^{2/3}), \qquad (4.6)$$

$$y(x) = y(x; \tau, p) = \begin{cases} 3^{1/2} \delta \beta^{3/2} (\beta - x)^{-1/2}, & x \in [0, b] \\ 3^{1/2}(\beta - x), & x \in [b, \beta], \end{cases} \qquad (4.7)$$

$$\Omega^{\pm} = \{z = x + iy : x \in \Delta^{\pm}, \ 0 < y < y(|x|)\}, \qquad \Omega = \Omega^+ + \Omega^-. \qquad (4.8)$$

Thus Ω^+ and Ω^- are two not intersecting domains symmetric with respect to imaginary axis. Next we note that the analytic continuations satisfy for $z \in D^{\pm}$

$$\mu'(-\bar{z}) = \mu'(\bar{z}), \qquad \Re\mu'(-\bar{z}) = \Re\mu'(\bar{z}); \qquad (4.9)$$

$$\phi(-\bar{z}) = \pi n - \phi(\bar{z}), \qquad \Im\phi(-\bar{z}) = \Im\phi(\bar{z}). \qquad (4.10)$$

Indeed μ' is even and real valued on Δ so (4.9) is valid for $z \in \Delta$. Hence the same holds for $z \in D^{\pm}$ by uniqueness theorem. Then we have $\phi(-x) = \pi n - \phi(x)$, $x \in \Delta$ (it follows by (4.2) since λ' is even and $|\mu'| = n + 1/2$) and (4.10) follows by the same theorem.

We say that the value of parameter p is admissible if the analytic continuation of ϕ from $\Delta^{\pm}$ to $\Omega = \Omega(p)$ satisfies

$$\Im\phi(z) < 0, \ z \in \Omega. \tag{4.11}$$

It follows by (4.10) that it is sufficient to check this condition only for $z \in \Omega^+$. We have by (4.2)

$$\Im\phi(x + iy) = -\pi \int_0^y \Re\mu'(x + it)dt \tag{4.12}$$

since $Re\mu'(x + it) > 0$, for $x \in [0, \beta), t = 0$. It follows by continuity that the same is true for $x \in (0, \beta), 0 < t < y_0(x)$ with some continuous $y_0(x) > 0$, $x \in [0, \beta)$. We shall show in what follows that $y(x; \beta, p) \leq y_0(x)$, $x \in [0, \beta]$, for any $p > 0$ and τ large enough.

For an admissible value p and $\Omega = \Omega(p)$, we define

$$\Lambda(z) = \log(1 + e^{-2i\phi(z)}), \ z \in \Omega. \tag{4.13}$$

where $\log \zeta$ stands for the branch in the right halfplane normalized by $\log 1 = 0$. By (4.11) this definition is correct.

Denoting by $\partial\Omega$ the boundary of a domain Ω we define

$$\Gamma^+ = \partial\Omega^+\backslash\Delta^+. \tag{4.14}$$

The curve Γ^+ consists of the three analytic arcs

$$\begin{aligned}
&\Gamma^+ = \Gamma_0^+ + \Gamma_1^+ + \Gamma_2^+, \text{ where} \\
&\Gamma_2^+ = \{iy : \ 0 \leq y \leq y(0)\}, \\
&\Gamma_0^+ = \{x + iy(x) : \ 0 \leq x \leq b\}, \\
&\Gamma_1^+ = \{\beta(1 - re^{-\pi i/3}) : \ 0 \leq r \leq 2\delta^{2/3}\}.
\end{aligned} \tag{4.15}$$

Similarly we define $\Gamma^- = \partial\Omega^-\backslash\Delta^- = \Gamma_0^- + \Gamma_1^- + \Gamma_2^-$ and finally

$$\Gamma = \Gamma^+ + \Gamma^-. \tag{4.16}$$

For any $k = 0, 1, 2$ the arcs Γ_k^+ and Γ_k^- are symmetric with respect to the imaginary axis. We note that Γ_2^+ and Γ_2^- being equal as sets will be distinguished to consider boundary values from Ω^+ on Γ_2^+ and boundary values from Ω^- on Γ_2^-.

THEOREM 3. *Let p be an admissible value and the path $\Gamma = \Gamma(p)$ has the orientation from β to $-\beta$. Then the following representations are true*

$$\eta(x) = \Im\frac{1}{\pi}\int_\Gamma \Lambda(t)\frac{dt}{x-t}, \qquad x \in (-\beta,\beta); \tag{i}$$

$$\theta(x) = \Im\frac{1}{\pi}\int_\Gamma \Lambda(t)\frac{dt}{x-t}, \qquad |x| > \beta; \tag{ii}$$

$$e_n = -\Re\frac{1}{\pi}\int_\Gamma e^{-2i\phi(t)}(\beta^2 - t^2)^{-1/2}dt. \tag{iii}$$

where e_n is defined by (2.10).

Proof. Here n is fixed and we drop the corresponding indecies. The function $T(z)/W(z)$ is analytic in $\bar{\mathbb{C}}\backslash\overset{\circ}{\Delta}$ and continuous on the boundary of this domain. By Lemma 3.1 and by definition we have

$$(T/W)^\pm(x) = 2e^{\eta(x)}\cos\phi(x)e^{\mp i\phi(x)}, \ x \in \Delta; \\ \Re(T/W)^\pm(x) = 2e^{\eta(x)}\cos^2\phi(x), x \in \Delta. \tag{4.17}$$

In particular $\Re(T/W)^\pm(x) \geq 0$, $x \in \Delta$. Hence $\Re(T/W)(z) \geq 0$ in the complement of Δ. This implies that we can define log in

$$\theta(z) = \log(T(z)/W(z)), \qquad \in \bar{\mathbb{C}}\backslash\Delta, \tag{4.2}$$

as the branch on the right half plane with normalization $\log 1 = 0$. The same branch of log was used in the definition of $\Lambda(z)$. sec (4.13). It follows by this remark that

$$-\pi/2 < \Im\theta(z) < \pi/2, \ z \in \bar{\mathbb{C}}\backslash\Delta \tag{4.18}$$

and the same with $\leq$ instead of $<$ holds for the boundary values. Taking this into account we obtain by (4.17)

$$\theta^\pm(x) = \eta(x) + \log|2\cos\phi(x)| \mp i\hat\phi(x) \tag{4.19}$$

$$\hat\phi(x) \equiv \phi(x)[\ \mod \pi], \ \hat\phi(x) \in [-\pi/2, \pi/2], x \in \Delta. \tag{4.20}$$

We see that (4.20) defines $\hat\phi(x)$ at all its continuity points and in particular for $x \in \Delta\backslash E$, where $E = \{x_1, ..., x_n, \pm\beta, 0\}$ is the set of zeros of T and the end points of $\Delta^\pm$.

Now we represent $\theta(z)$, $z \in \bar{\mathbb{C}}\backslash\Delta$ as the Cauchy integral over the boundary of the domain. This yields by (4.19)

$$\theta(z) = \frac{1}{2\pi i}\int_\Delta (\theta^-(t) - \theta^+(t))\frac{dt}{t-z} = -\frac{1}{\pi}\int_\Delta \hat\phi(t)\frac{dt}{t-z} \tag{4.21}$$

Now (4.19) and (4.21) yield the two following representations

$$\Re\theta^+(x) = \eta(x) + \log | 2 \cos \phi(x) | = -\frac{1}{\pi} \int_\Delta \hat{\phi}(t) \frac{dt}{t-x}, \qquad (4.22)$$

where the second one follows by Plemelj formula (3.1).

Further, it follows from our remarks about the branch of log in (4.13) that (4.18) is true with $\Lambda(z)$ replaced by $\theta(z)$. On the other hand we can rewrite that definition as

$$\Lambda(z) = \log(2 \cos \phi(z)) - i\phi(z), \ z \in \Omega \qquad (4.23)$$

and we obtain for the boundary values of Λ

$$\Lambda(x) = \log | 2 \cos \phi(x) | -i\hat{\phi}(x), \ x \in \Delta \backslash E, \qquad (4.24)$$

where $\hat{\phi}(x)$ is defined by (4.20). We have by (4.24) and (4.19)

$$\Im\theta^+(x) = \Im\Lambda(x) = -\hat{\phi}(x), \ x \in \Delta \backslash E. \qquad (4.25)$$

Using this we can rewrite (4.22) as

$$\eta(x) = -\log | 2 \cos \phi(x) | +\Im\frac{1}{\pi} \int_\Delta \Lambda(t) \frac{dt}{t-x}, \ x \in \Delta \backslash E. \qquad (4.26)$$

Now let $x \in \Delta \backslash E$ be fixed and $C_\epsilon(x) = \{z = x + \epsilon e^{it} : 0 \leq t \leq \pi/2\}$ be the half circle with the orientation from $x + \epsilon$ to $x - \epsilon$. We have

$$\lim_{\epsilon \to 0} \Im\frac{1}{\pi} \int_{C_\epsilon} \Lambda(t) \frac{dt}{t-x} = \Im(i\Lambda(x)) = \log | 2 \cos \phi(x) | \qquad (4.27)$$

(the limit equals to the half of the respective residue; the second equality follows by (4.24)). On the other hand the integral on the right hand side of (4.26) is the principal value, that is it equals to the limit as $\epsilon \to 0$ of that integral over $\Delta \backslash (x - \epsilon, x + \epsilon)$ from the left to the right. Taking this into account we add (4.26) to (4.27) and then change the orientation of C_ϵ to the opposite; this yields

$$\eta(x) = \lim_{\epsilon \to 0} \Im\frac{1}{\pi} \int_{\Gamma_\epsilon} \Lambda(t) \frac{dt}{x-t}, \ x \in \Delta \backslash E, \qquad (4.28)$$

where the orientation on $\Gamma_\epsilon = [\beta, x + \epsilon] + C_\epsilon + [x - \epsilon, -\beta]$ is taken from β to $-\beta$. But the integral over Γ_ϵ above does not depend on ϵ by Cauchy theorem provided ϵ is small enough. Thus we can omit the sign of lim on the right hand side of (4.28). But now the integral over Γ_ϵ equals to that over Γ by the same theorem. We have proved (i) for $x \in \Delta \backslash E$. It remains to notice that the assertion for $x \in (-\beta, \beta)$ follows by continuity.

Next, we substitute ϕ in (4.21) with its expression by (4.25). This yields

$$\theta(x) = \frac{1}{\pi} \int_\Delta \Im \Lambda(t) \frac{dt}{t-x} = \Im \int_\beta^{-\beta} \Lambda(t) \frac{dt}{x-t}$$

The last two integral over Δ are equal to those over Γ by Cauchy theorem and we have proved (ii).

Finally using the definition of e_n (10.2) we substitute $\cos 2\phi$ in it with $\Re e^{-2i\phi}$ to obtain

$$e_n = \Re \frac{1}{\pi} \int_\Delta e^{-2i\phi(t)} (\beta^2 - t^2)^{-1/2} dt.$$

Now we apply again Cauchy integral theorem to substitute here Δ with Γ and obtain (iii). The proof is completed.

Remark. It is clear from the proof that the statement of the theorem does not require any special form of Γ. In fact we have proved that given x these assertions are valid for any curve Γ which is homotopic to $\Gamma_\epsilon(x)$ (see above) with respect to Λ (in particular we need analytisity Λ at x only; moreover, we have also obtained representations for nonanalytic μ').

Next, we make use of the symmetry with respect to imaginary axis.

It is convenient now to introduce special notation for the pieces of the integral representation (i) using the structure of Γ. We define

$$\eta_i^\pm(x) = \Im \frac{1}{\pi} \int_{\Gamma_i^\pm} \Lambda(t) \frac{dt}{x-t}, \quad i = 0,1,2; \tag{4.29}$$

$$\eta_i(x) = \eta_i^+(x) + \eta_i^-(x), \quad i = 0,1,2. \tag{4.30}$$

The assertion (i) of Theorem 1 can now be presented as

$$\eta(x) = \eta_0(x) + \eta_1(x) + \eta_2(x). \tag{4.31}$$

We want to notice in addition that for any $i = 0,1,2$ we have

$$\eta_i^-(x) = \eta_i^+(-x), \quad x \in (-\beta, \beta). \tag{4.32}$$

Indeed we deduce from (4.10) that $\Lambda(-\bar{z}) = \overline{\Lambda(z)}$, $z \in \Gamma$. Taking this into account we make substitution $t = -\zeta$ in the integral (4.29) over Γ_i^+ defining η_i^+ to obtain

$$\eta_i^-(x) = \Im \frac{1}{\pi} \int_{\Gamma_i^-} \Lambda(t) \frac{dt}{x-t} = \Im \frac{1}{\pi} \int_{\Gamma_i^+} \Lambda(-t) \frac{dt}{x+t} = \Im \frac{1}{\pi} \overline{\int_{\Gamma_i^+} \Lambda(t) \frac{dt}{x+t}}$$

(note that our substitution changes the orientation); the last integral equals to $\eta_i^+(-x)$ and (4.32) follows.

Finally we take into account the special form of Γ and of the function $\nu(x) = \alpha x^\rho$, $\rho > 1$ to prove the following

LEMMA 4.1. For sufficiently large τ satisfying (2.1) and $x \in [-1, 1]$ we have

$$| \eta_0(\beta x) | \leq n^{-Cp}, \qquad x \in \mathbb{R}, \tag{i}$$

$$|\eta_2(\beta x)| \leq C_2 \frac{n^{1-\rho}}{1 + n^2 x^2}, \quad x \in \mathbb{R}, \tag{ii}$$

$$-\frac{C_1}{n^{2/3}(1 - |x|)} \leq \eta_1(\beta x) \leq \frac{C_1}{1 + n^{2/3}(1 - |x|)}, \quad x \in \mathbb{R}, \tag{iii}$$

$$|\theta(x)| \leq C_3, \ x \in \mathbb{R} \backslash \Delta, \tag{iv}$$

where C, C_i are positive constants which may depend only on ρ.

Proof. To simplify our calculations we suppose that

$$\tau = n + 1/2$$

General case $0 \leq n + 1/2 - \tau \leq M$ requires only few obvious modifications. Using notation

$$\lambda^1(z) = \pi^{-1} \int_x^\beta \nu'(t)(t^2 - z^2)^{-1/2} dt, \tag{4.33}$$

$$\lambda^0(z) = \pi^{-1} \int_z^x \nu'(t)(t^2 - z^2)^{-1/2} dt, \tag{4.34}$$

we have by (4.2)

$$\mu_n' = \lambda'(z) = \lambda^0(z) + \lambda^1(z). \tag{4.35}$$

We also denote $\vartheta(\xi, z) = \arg(\xi^2 - z^2)^{-1/2} \nu'(\xi)$ and obtain

$$0 \leq \vartheta(t, z) = \frac{1}{2} \arctan \frac{y}{t - x} + \frac{1}{2} \arctan \frac{y}{t + x} \leq \pi/4$$

$$| t^2 - z^2 |^2 = (t^2 - x^2 + y^2)^2 + 4x^2 y^2 \leq 37\beta^2(\beta - x)^2, \tag{4.36}$$

where $t \in [x, \beta)$, $z = x + iy \in \Omega^+$. Hence

$$\Re\lambda^1(z) = \pi^{-1} \Re \int_x^\beta \nu'(t)(t^2 - z^2)^{-1/2} dt \geq C_5 (\frac{\beta - x}{\beta})^{1/2} \frac{\nu(\beta) - \nu(x)}{\beta - x}$$

The last factor is not less then $\nu(\beta)/\beta$ by the convexity ν and we obtain

$$\Re\lambda^1(z) \geq C_6 \beta^{-3/2}(\beta - x)^{1/2} n. \tag{4.37}$$

Next, for $\xi = x + it$, $z = x + iy \in \Omega^+$ we have

$$0 \leq \vartheta(\zeta, z) = \pi/4 + (\rho - 1) \arctan \frac{t}{x} - \frac{1}{2} \arctan \frac{t + y}{2x} \leq$$

$$\leq \pi/4 + (\rho - 1) \arctan \frac{y(x)}{x}.$$

For $x \geq 1$ and sufficiently large n we have $0 \leq \theta(\zeta, z) \leq \pi$ and

$$\pi \Re \lambda^0(z) = \Im \int_0^y \nu(\zeta)(\zeta^2 - z^2)^{-1/2} dt \geq 0$$

Hence, (4.37) remains valid with $\lambda'(z)$ instead of $\lambda^1(z)$.
For $x \in [0, 1]$

$$\pi \mid \lambda^0(z) \mid \leq \int_0^y \mid \nu'(x + it) \mid (y - t)^{-1/2} (4x^2 + (y + t)^2)^{-1/4} dt \leq C_8$$

and again (4.37) remains valid for $\lambda'(z)$ with another constant. Hence

$$\Re \lambda'(x + iy) \geq C_9 \beta^{-3/2} (\beta - x)^{1/2} n, \quad x \in \Delta, \ 0 \leq y \leq y(x) \qquad (4.39)$$

Using (4.5-7), (4.12) and (4.39) we obtain

$$\mid \exp\{-2i\phi(x + iy(x))\} \mid \leq n^{-Cp}, x \in \Delta.$$

The same is true for $\Lambda(x + iy(x))$ with extra $\log n$ and (i) follows.
Next, we have

$$\lambda^1(iy) = \pi^{-1} \int_0^\beta \nu'(t)(t^2 + y^2)^{-1/2} dt = \rho \alpha_\rho \pi^{-1} \int_0^\beta t^{\rho - 1} dt (1 - \epsilon_3) =$$
$$= \pi^{-1} \rho(\rho - 1)^{-1} \tau \beta^{-1}(1 - \epsilon_3),$$

where $0 \leq \epsilon_3 = \epsilon_3(y) = O((y/\beta)^{\rho - 1})$ as $y \to 0$.

$$\lambda^0(z) = -\pi^{-1} \int_0^y \nu'(it)(y^2 - t^2)^{-1/2} dt =$$
$$= -\rho \alpha_\rho \pi^{-1} (iy)^{\rho - 1} \int_0^{y(0)} t^{\rho - 1} (y^2 - t^2)^{-1/2} dt = \rho(2\pi)^{-1}(iy)^{\rho - 1}.$$

These two relations together imply successively

$$\lambda'(iy) = \pi^{-1} \rho(\rho - 1)^{-1} \tau \beta^{-1}(1 - \epsilon_3) + \rho(2\pi)^{-1}(iy)^{\rho - 1};$$

$$-2i\phi(iy) = -\pi i n + (-2\pi i)(-i) \int_0^y \lambda'(it) dt =$$
$$= -\pi i n - \rho(\rho - 1)^{-1} \tau \beta^{-1} y(1 - \epsilon_3) + i(iy)^\rho;$$

$$\Im \Lambda(iy) = \arg(1 + e^{-2i\phi(iy)}) =$$
$$= \arg(1 + (-1)^n \exp\{-\rho(\rho - 1)^{-1} \tau \beta^{-1} y(1 - \epsilon) + i(iy)^\rho\}) . \qquad (4.40)$$

On the other hand we have by (4.33) and (4.34)

$$\eta_2(x) = \eta_2^+(x) + \eta_2^+(-x) = \pi^{-1}\Im\int_{\Gamma_2^+}\Lambda(\zeta)\{(-x-\zeta)^{-1} + (x-\zeta)^{-1}\}d\zeta =$$

$$= \pi^{-1}\int_{\Gamma_2^+}\Im\Lambda(\zeta)(x^2-\zeta^2)^{-1}d\zeta^2 = \pi^{-1}\int_0^{y(0)}\Im\Lambda(iy)(x^2+t^2)^{-1}dy^2.$$

We substitute here $\Im\Lambda(iy)$ with its expression by (4.40), then substitute x with βx and finally make the substitution $t = \tau\beta\nu(\beta)^{-1}$ in the integral. After straightforward transformations this yields

$$\pi\eta_2(\beta x) = \tau^{1-\rho}\int_0^{p\sqrt{3}\log\tau} A_n(t)\frac{dt^2}{t^2+\tau^2x^2}, \qquad x\in\mathbb{R};$$

where

$$A_n(t) = s_n c\frac{t^\rho e^{-Mt+\epsilon(t)}}{1+s_n e^{-Mt+\epsilon(t)}},$$

$s_n = (-1)^n$, $M = \rho/(\rho-1)$, $c = \cos(\pi\rho/2)$ and $\epsilon(t) = |\epsilon_n(t)| \le Ct^\rho n^{1-\rho}$ for $t\in[0, p\sqrt{3}\log\nu(\beta)]$. Now (ii) follows by this representation.

For $\zeta\in\Gamma_1^+$ we have by (4.15)

$$\zeta = \beta(1-\bar\sigma r), \ r\in[0, 2\sigma^{2/3}], \ \sigma = e^{\pi i/3},$$

$$\Phi(\zeta) = \int_\zeta^\beta\int_z^\beta\nu'(t)\frac{dtdz}{\sqrt{t^2-z^2}} - \frac{\pi}{4} =$$

$$= \frac{2}{3}\sqrt{2}\beta\nu'(\beta)(1-\frac{\zeta}{\beta})^{\frac{3}{2}}(1+O(1-\frac{\zeta}{\beta})) - \frac{\pi}{4} = \qquad (4.41)$$

$$= -i\frac{2}{3}\sqrt{2}\tau r^{3/2}(1+O(r)) - \frac{\pi}{4}$$

and therefore

$$\Lambda(\zeta) = \log(1+ie^{-2i\Phi(\zeta)}) = \log(1+ie^{-M\tau r^{3/2}} + \epsilon_1(r))$$

$$= \log(1+ie^{-M\tau r^{3/2}}) + \epsilon_2(r),$$

where $\epsilon_1(r)$, $\epsilon_2(r) = O(\tau r^{5/2}e^{-M\tau r^{3/2}})$.
Using $t = \tau^{2/3}r$ we obtain

$$\Lambda(\zeta) = \Lambda(\zeta(t)) = \log(1+ie^{-M\tau t^{3/2}}) + \epsilon_3(t)$$

$$= A_2(t) + iA_1(t) + \epsilon_3(t) \qquad (4.42)$$

where

$$A_2(t) = \frac{1}{2}\log(1 + e^{-2Mt^{3/2}}),$$

$$A_1(t) = \arctan(e^{-Mt^{3/2}}),$$

$$\epsilon_3(t) = O(e^{-Mt^{3/2}}\tau^{-2/3}t^{5/2}).$$

Next we have for $\zeta \in \Gamma_1^+$

$$\frac{d\zeta}{x - \zeta} = \frac{\bar{\sigma}dr}{(1 - \frac{x}{\beta}) - \bar{\sigma}r} = \frac{\bar{\sigma}dt}{y_n - \bar{\sigma}t} =$$

$$= \frac{1}{2}(\sqrt{3} - i)\frac{y_n dt}{|\, y_n - \bar{\sigma}t\,|^2} - \frac{tdt}{|\, y_n - \bar{\sigma}t\,|^2}$$

(4.43)

where $y_n = \tau^{2/3}$, $\tau = n + 1/2$, $\beta = \beta_\tau$.

It follows by (4.42), (4.43)

$$\pi\eta_1^+(x) = \Im\int_{\Gamma_1^+}\Lambda(\zeta)\frac{d\zeta}{x - \zeta} = \Im\int_0^R\Lambda(\zeta(t))\frac{\bar{\sigma}dt}{y_n - \bar{\sigma}t} =$$

$$\overset{\cdot}{=} \epsilon_{1,n}(x) - \epsilon_{2,n}(x) + \epsilon_{3,n}(x),$$

where $R = R_n = 2(p\log\tau)^{2/3}$ and

$$\epsilon_{1,n} = \frac{1}{2}\int_0^R(\sqrt{3}A_1(t) - A_2(t))\frac{y_n dt}{|\, y_n - \bar{\sigma}t\,|^2},$$

$$\epsilon_{2,n} = \frac{1}{2}\int_0^R A_1(t)\frac{tdt}{|\, y_n - \bar{\sigma}t\,|^2},$$

$$\epsilon_{3,n} = O(\tau^{-2/3}\int_0^R t^{5/2}e^{-Mt^{3/2}}\frac{tdt}{|\, y_n - \bar{\sigma}t\,|}) = O(n^{-2/3}\frac{1}{1 + |\, y_n\,|})$$

(note that $|\, y_n - \bar{\sigma}t\,|^2 = y_n^2 - ty_n + t^2 \geq \frac{1}{2}(y_n^2 + t^2)$). We have from these representations

$$\epsilon_{1,n}(x) = O(\frac{1}{1 + |\, y_n\,|}), \qquad x \in \mathbb{R},$$

$$0 \leq \epsilon_{2,n}(x) = O(\frac{1}{|\, y_n\,|}), \qquad x \in \mathbb{R},$$

and $\epsilon_{3,n}$ is small with respect to $\epsilon_{1,n}$.

By (4.32) the same is true for $\eta_1(x)$ with

$$y_n = (1 - \frac{|\, x\,|}{\beta})(n + 1/2)^{2/3},$$

and (iii) follows. Assertion (iv) follows by (i–iii) of Theorem 3. Lemma 4.1 is proved.

Now we can prove assertions of Lemma 4.

First, we have by (4.1)

$$| T_n(x) - R_n(x) | = | R_n(x) | | e^{\eta_n(x)} - 1 |$$

and (i) of Lemma 4 follows by (2.6), (2.7) and Lemma 4.1.

Next (iv) follows by (iii) of Theorem 3 and the estimates for Φ_n which were obtained in the proof of Lemma 4.1.

For $\tau = \tau_n$ satisfying (2.1) we have from (3.2), (2.2)–(2.4) and (iv) of Lemma 1

$$| W_n(x) |^2 h(x) = \frac{1}{2\pi}(x^2 - \beta_\tau^2)^{-\frac{1}{2}} e^{-2(V^{\lambda\tau} + \phi - w_\tau)(x) + (n + \frac{1}{2} - \tau)g_\tau(x)}$$

$$< (\frac{2x}{\beta_\tau})^M (x^2 - \beta_\tau^2)^{-\frac{1}{2}} e^{-Cn\beta_\tau^{-1}|x|^{-1/2}(|x| - \beta_\tau)^{3/2}}, \ | x | \ge \beta_\tau.$$

Hence

$$\int_{\mathbb{R}\backslash\Delta_\tau} | W_n(x) |^2 h(x)dx = O(n^{-1/3})$$

and the same is true for T_n by (iv) of Lemma 4.1.

It proves (ii) of Lemma 4.

Finally we have

$$\| T_n \|_\tau^2 = \| T_n - R_n \|_\tau^2 + 2(R_n, T_n)_\tau - \| R_n \|_\tau^2 =$$
$$= \| T_n - R_n \|_\tau^2 + 1 - e_n$$

(see (3.4) and Lemma 3.3). Now (iii) follows by (ii), (iv) and (2.15). The proof is completed.

5. An upper bound for the Szegö kernel function

Let $n \in \mathbf{N}, \tau > 0$. We shall use the notations:

$$K_n(x, y) = K_n(x, y; \tau) = \sum_{k=1}^{n} Q_k(x, \tau)Q_k(y, \tau); \ K_n(z) = K_n(z, \bar{z}).$$

We need also Cristoffel-Darboux formula

$$K_n(x, y) = (x - y)^{-1}(Q_{n+1}(x)Q_n(y) - Q_n(x)Q_{n+1}(y)). \tag{5.1}$$

and the well-known extremal property

$$K_n(z) = \max_{q \in \mathbb{P}_n}(| q(z) |^2 / q_\tau^2). \tag{5.2}$$

All the required estimates for K_n are immediate corollaries of the following simple auxiliary lemma (see [10]).

LEMMA 5.1. *For any* $n \in \mathbb{N}$, $\tau \leq n + 1/2$ *we have with* $m = n + 1/2$

$$K_n(z, \tau) \leq \pi^{-1} \mathrm{dist}^{-1}(z, \Delta) \exp\{w_{m,\tau} - V^{\mu_{m,\tau}}(z)\}, \quad z \in \mathbb{C} \backslash \Delta_\tau.$$

Proof. Denoting $W(z) = \exp\{w_{m,\tau} - U^{\mu_{m,\tau}}(z)\}$, we consider for any $q \in \mathbb{P}_n$ the function $f(z) = q(z)/W(z)$, which is holomorphic in $z \in \mathbb{C} \backslash \Delta_\tau$ including ∞ and satisfies $f(\infty) = 0$. Representing f^2 in this domain by the Cauchy integral and then making the rough estimate of its modulus, we obtain

$$\mid f(z) \mid^2 \leq \pi^{-1} \mathrm{dist}^{-1}(z, \Delta) \int_{\Delta_\tau} \mid q(t) \mid^2 \mid W(t) \mid^{-2} dt, \quad z \in \mathbb{C} \backslash \Delta.$$

Taking into account the boundary value of W(see definitions and the equilibrium conditions for $\mu_{m,\tau}$ in Lemma 3.1) we obtain that $|W(t)|^{-2} = h(t)$, hence the integral on the right hand side equals to q_τ. The assertion of lemma follows by (2.6) and the definition of f.

LEMMA 5.2. *For any* $n \in \mathbb{N}$, $M \geq n - \tau + 1/2 \geq 0$ *we have*

$$h(x)K_n(x) \leq C(M)n/\beta_\tau, \qquad x \in \mathbb{R}$$

Proof. Let $y > 0$. It follows from the definition of K_n that $K_n(x) \leq K_n(x + iy), x \in \mathbb{R}$. Now we put $y = \beta_\tau/n$ and use Lemma 5.1. Taking into account equilibrium conditions for $\mu_{m,\tau}$ we obtain

$$h(x)K_n(x) \leq n\beta_\tau^{-1} \exp\{V^{\mu_{m,\tau}}(x) - V^{\mu_{m,\tau}}(x + iy)\}$$
$$\leq n\beta_\tau^{-1} \exp\{V^{\lambda_\tau}(x) - V^{\lambda_\tau}(x + iy) + pg_\tau(x)\}.$$

where $p = n - \tau + 1/2 \in [0, M]$. Then we have by (1.16), (1.17)

$$V^{\lambda_\tau}(x) - V^{\lambda_\tau}(x + iy) = 1/2 \int_{\Delta_\tau} \log(1 + y^2(x - t)^{-2})d\lambda_\tau(x) =$$

$$Cn \int_{-1}^{1} \log(1 + n^{-2}(x/\beta_\tau - t)^{-2})v(t)dt \leq$$

$$\leq C_1 n \int_{-1}^{1} \log(1 + n^{-2}(x/\beta_\tau - t)^{-2})dt \leq$$

$$C_1 n \int_{-\infty}^{\infty} \log(1 + n^{-2}(x/\beta_\tau - t)^{-2})dt = C_1 n \int_{-\infty}^{\infty} \log(1 + n^{-2}t^{-2})dt = C_1.$$

(we use also that $v(x)$ is bounded in $[-1, 1]$). It remains to notice that $g_\tau(x + i\beta_\tau/n) = O(1)$ under our conditions. Lemma is proved.

LEMMA 5.3. *Let* $q \in \mathbb{P}_n$, $0 \leq n - 1/2 + \tau \leq M$ *and*

$$\int_{\Delta_\tau} | q(x) |^2 h(x)^2 dx \leq 1$$

Then

$$\int_{\mathbb{R} \backslash \Delta_\tau} | q(x) |^2 h(x)^2 dx \leq C \log n, \quad where \quad C = C(M).$$

Proof. The assertion follows easily by the two previous lemmas and Lemma 1, (iv).

6. Proof of Theorem 1 and 2, assertions (iii)

1. Proof of Theorem 2, (i)

In this section we consider only $\tau_n = n + 1/2$ and use the special notation for this case:

$$\Delta(n) = \Delta_{n+1/2}, \ \Delta(n, \epsilon) = [-(1 - \epsilon)\beta_{n+1/2}, (1 - \epsilon)\beta_{n+1/2}],$$
$$K_{n+j}(x, y) = K_{n+j}(x, y; n + 1/2), \ K_{n+j}(x) = K_{n+j}(x; n + 1/2),$$
$$Q_{n+j}(x) = Q_{n+j}(x, n + 1/2) \text{ for } n \in \mathbb{N}, \ j = 0, 1, 2, \dots.$$

LEMMA 6.1. *Let* $\epsilon \in (0, 1/2)$, $M \in \mathbb{N}$ *be fixed;* $m \in [0, M]$ *be a nonnegative integer and* $s = \min\{1/6, \rho - 1\}$. *Then we have*

$$Q_{n+m}(x) = (1 + O(n^{-s}))T_{n+m}(x) + O(n^{-s})Q_{n+m+1}(x) + \epsilon_1(x, n, m) \tag{i}$$

$$Q_{n+m+1}(x) = (1 + O(n^{-s}))T_{n+m+1}(x) + O(n^{-s})Q_{n+m}(x) + \epsilon_2(x, n, m) \tag{ii}$$

where

$$\epsilon_i(x, n) = O(n^{-s}\beta_{n+1/2}^{-1/2}h(x)^{-1/2}), \ i = 1, 2.$$

uniformly with respect to $x \in \Delta(n, 2\epsilon)$ *and* $m \in [0, M]$.

Proof. Using Lemma 3, (i') and (5.1) we obtain

$$Q_{n+m}(x) = u_{n,m} \int_{\Delta(n)} R_{n+m}(t)K_{n+m}(t, x)h(t)dt =$$

$$u_{n,m}T_{n+m}(x) + \int_{\Delta(n)} (R_{n+m}(t) - T_{n+m}(t))K_{n+m}(t, x)h(t)dt = \tag{6.1}$$

$$= u_{n,m}T_{n+m}(x) + \epsilon_{n,1}(x) + \epsilon_{n,2}(x)Q_{n+m}(x) + \epsilon_{n,3}(x)Q_{n+m+1}(x),$$

where $u_{n,m} = c_{n+m}^{-1} b_{n+m}$ and (we don't indicate dependence on m)

$$\epsilon_{n,1}(x) = \int_{\Delta(n,\epsilon)} (R_{n+m}(t) - T_{n+m}(t)) K_{n+m}(t,x) h(t) dt,$$

$$\epsilon_{n,2}(x) = b_{n+m+1}^{-1} b_{n+m} \int_{\Delta \backslash \Delta(n)} (R_{n+m}(t) - T_{n+m}(t)) Q_{n+m+1}(t) \frac{h(t) dt}{x-t}$$

$$\epsilon_{n,3}(x) = b_{n+m+1}^{-1} b_{n+m} \int_{\Delta \backslash \Delta(n)} (R_{n+m}(t) - T_{n+m}(t)) Q_{n+m}(t) \frac{h(t) dt}{x-t}.$$

In addition to this representation we have by Theorem 2, (i)

$$u_n = 1 + O(n^{-1/3}),$$
$$b_{n+m+1}^{-1} b_{n+m} = 2^{-1} \beta_{n+1/2} (1 + O(n^{-1/3})). \tag{6.2}$$

We have also by the definition of $\delta_n(y)$ (see Lemma 4)

$$\beta_{n+1/2} \int_{\Delta(n,\epsilon)} \delta_n^2(\beta_{n+1/2} \, x) dx = \int_{1-\epsilon}^{1+\epsilon} \delta_n^2(y) dy = O(n^{-1-2s});$$

Taking all these into account we use the Cauchy inequality and then Lemma 4, (i) and Lemma 5.2 to obtain by (1.7)

$$\epsilon_{n,1}(x) = O(n^{-s} \beta_{n+1/2}^{-1/2} h(x)^{-1/2}) \text{ and } \epsilon_{n,i}(x) = O(n^{-s}), \ i = 2,3$$

as $n \to \infty$ uniformly with respect to $x \in \Delta(n,\epsilon)$. It remains to move $\epsilon_{n,2}(x) Q_{n+m}(x)$ to the left hand side and to divide by $1 - \epsilon_2(x)$; (i) follows. To prove (ii) we substitute $K_{n+m}(x,y) = K_{n+m-1}(x,y) - Q_{n+m}(x) Q_{n+m}(y)$ in (2.6) (the second equality) and further do the same.

To complete the proof we solve the system (i)–(ii) of linear equations with respect to "unknown" functions $Q_{n+m}(x), Q_{n+m+1}(x)$ and obtain

$$Q_{n+m}(x) = (1 + O(n^{-s})) T_{n+m}(x) + \epsilon(x,n,m) \tag{6.3}$$

where

$$\epsilon(x,n,m) = O(n^{-s} \beta_{n+1/2}^{-1/2} h(x)^{-1/2}), \ i = 1,2.$$

uniformly with respect to $x \in \Delta(n,2\epsilon)$ and $m \in [0,M]$. Theorem 2,(iii) follows by Lemma 4, (i).

2. Proof of Theorem 1, (iii)

LEMMA 6.2. Let $\epsilon \in (0, 1/2)$, $M \in \mathbf{N}$ be fixed; $m \in [0, M]$ be a nonnegative integer and $s = \min\{1/6, \rho - 1\}$. Then we have

$$H_{n+m}(x) = (1 + \epsilon_{1,n})Q_{n+m}(x) + \epsilon_{2,n}Q_{n+m-1}(x)$$

where $\epsilon_{n,i} = O(n^{-s})$ as $n \to \infty$ uniformly with respect to $x \in \Delta(n, 2\epsilon)$ and $m \in [0, M]$.

Proof. Using the Koraus identity we obtain

$$H_{n+m}(x) - b_{n+m}^{-1}a_{n+m}Q_{n+m}(x) =$$

$$= \int_{\Delta(n)} H_{n+m}(t)K_{n+m-1}(t, x)h(t)dt =$$

$$= \int_{\mathbb{R}\backslash\Delta(n)} H_{n+m}(t)K_{n+m-1}(t, x)h(t)dt =$$

$$= \epsilon_{n,1}(x)Q_{n+m}(x) + \epsilon_{n,2}(x)Q_{n+m+1}(x)$$

where

$$\epsilon_{n,1}(x) = b_{n+m+1}^{-1}b_{n+m}\int_{\Delta\backslash\Delta(n)} H_{n+m}(t)Q_{n+m-1}(t)\frac{h(t)dt}{x-t},$$

$$\epsilon_{n,2}(x) = b_{n+m+1}^{-1}b_{n+m}\int_{\Delta\backslash\Delta(n)} H_{n+m}(t)Q_{n+m}(t)\frac{h(t)dt}{x-t}.$$

Using the Cauchy inequality, (6.2) and Lemma 6.3 we obtain

$$\epsilon_{n,i}(x) = O(n^{-s}\log n), \quad i = 1, 2$$

and lemma follows.

Theorem 1, (iii) follows by Theorem 2, (iii).

References

[1] V.S.Bujarov, *On Logarithmic Asymptotics for Polynomials Orthogonal on* $\mathbb{R}$ *with non-symmetric weights*, Mat. Zametki 50 no. 2 (1991), pp. 28–36. (In Russian)

[2] A.A.Gonchar and E.A.Rakhmanov, *Equilibrium measure and the distribution of Zeros of Extremal Polynomials*, Math. USSR Sbornik **53** (1986), pp. 119–130; Transl. from Russian **125(167)** no. 1 (1984).

[3] G.Lopez and E.A.Rakhmanov, *Rational Approximations, Orthogonal Polynomials and Equilibrium distributions*, Lecture Notes in Math. **1329** (1988); Proceedings of Segovia (1986.).

[4] D.S.Lubinsky and E.B.Saff, *Strong Asymptotics for Extremal Polynomials Associated with Weights on* $\mathbb{R}$, Lecture Notes in Math. **1305** (1986).

[5] D.S.Lubinsky, *Strong Asymptotics for Extremal Errors and Polynomials Associated with Erdös-type Weights*, Pitman Reseach Notes in Mathematics Series **202** (1989).

[6] P.Nevai, *Asymptotics for Orthogonal Polynomials Associated with* $\exp(-x^4)$, SIAM.J. Math .Anal. **15** (1984), 1177–1187.

[7] P.Nevai and Gesa Freud, *Orthogonal Polynomials and Cristoffel Functions, A Case Study*, J. Approx. Theory **48** (1986), pp. 3–167.

[8] N.H.Mhaskar and E.B.Saff, *Extremal Problems for Polynomials with Exponential Weights*, Trans. Amer. Math. Soc. **286** (1984), pp. 203–234.

[9] E.A.Rakhmanov, *On Asymptotic Properties of Polynomials Orthogonal on the Real Axis*, Dokladi of Acad.of Sci. of USSR **261** No2 (1981), pp. 282–284. (In Russian)

[10] E.A.Rakhmanov, *On Asymptotic Properties of Polynomials Orthogonal on the Real Axis*, Math. USSR Sb. **47** (1984), pp. 155-193; Transl. from Russian **119** (**161**) (1982), pp. 163–203.

[11] E.A.Rakhmanov, *"Asymptotic Properties of Orthogonal Polynomials"*, *Doctoral Thesis,*, Steklov Math Inst, Moscow, 1983.

[12] R.C.Sheen, *"Orthogonal Polynomials Associated with* $\exp(-x^6/6)$*".*, *Ph.D. dissertation*, Ohio State University, Columbus, 1984.

Steklov Math.Inst.
Dept. of Theory of Function of Complex Variable
Moscow, Vavilova, 42
Russia

EXACT CONVERGENCE RATES FOR BEST L_P RATIONAL APPROXIMATION TO THE SIGNUM FUNCTION AND FOR OPTIMAL QUADRATURE IN H^P

A.L. Levin[1] and E.B. Saff[2]

For the signum function s on $[-1, 1]$, the exact rate of convergence (up to a multiplicative constant) for best L_p approximation to s by real rational functions of degree n was obtained by Vjacheslavov for $1 \leq p < \infty$. Here we show that the same rates hold for complex rational approximation and we obtain extensions to L_p for $0 < p < \infty$. Exact L_p convergence rates are also obtained for best symmetric rational approximation on the unit circle to the function $H_\omega(z)$ that equals one on the arc $z = e^{i\theta}$, $-\omega < \theta < \omega$, and zero elsewhere. These results yield a new proof for the exact convergence rate for optimal quadrature in the Hardy space H^p.

1. Introduction

$$\text{Let} \quad s(x) := \begin{cases} 1, & 0 < x \leq 1 \\ -1, & -1 \leq x < 0 \end{cases}$$

denote the signum function on $[-1, 1]$ and, for $0 < p < \infty$, set

$$\delta_{n,p}(s) := \inf_{r_n \in \mathcal{R}_n} \|s - r_n\|_{L_p[-1,1]} = \inf_{r_n \in \mathcal{R}_n} \left(\int_{-1}^{1} |s(x) - r_n(x)|^p dx \right)^{1/p}, \quad (1.1)$$

where $\mathcal{R}_n$ denotes the collection of all rational functions with complex coefficients having degree at most n (i.e., $r_n \in \mathcal{R}_n$ iff $r_n = P/Q$, $\deg P \leq n$, $\deg Q \leq n$).

---

[1] Research conducted while visiting the University of South Florida.

[2] Research supported in part by the National Science Foundation under grant DMS-881-4026.

Vjacheslavov [12] constructed (for any $p > 0$) a sequence of rational functions $r_n^* \in \mathcal{R}_n$ with real coefficients such that

$$\|s - r_n^*\|_{L_p[-1,1]} \leq C_p n^{1/2p} e^{-\pi\sqrt{n/p}}, \quad n = 1, 2, \dots .$$

Here C_p depends only on p and may be replaced by an absolute constant provided $p \geq 1$.

Vjacheslavov also showed that for any *real-valued* $r_n \in \mathcal{R}_n$ and for $p \geq 1$, the following inequality holds:

$$\|s - r_n\|_{L_p[-1,1]} \geq c_0 n^{1/2p} e^{-\pi\sqrt{n/p}}, \quad n = 1, 2, \dots , \tag{1.2}$$

where $c_0 > 0$ is an absolute constant. One aim of this paper is to show that (1.2) holds (with some $c_p > 0$ instead of c_0) for any complex-valued $r_n \in \mathcal{R}_n$ and for any $p > 0$. We thus obtain

$$c_p n^{1/2p} e^{-\pi\sqrt{n/p}} \leq \delta_{n,p}(s) \leq C_p n^{1/2p} e^{-\pi\sqrt{n/p}}, \quad n = 1, 2, \dots . \tag{1.3}$$

For problems in signal processing a more appropriate setting for rational approximation is on the unit circle $C : |z| = 1$ of the complex plane $\mathbf{C}$. For $0 < \omega < \pi$, we consider the ideal filter

$$H_\omega(z) := \begin{cases} 1, & \text{if } z = e^{i\theta}, \quad -\omega < \theta < \omega \\ 0, & \text{otherwise} \end{cases}. \tag{1.4}$$

Since $H_\omega(z) = H_\omega(1/z)$ for $z \in C$, it is natural to restrict our attention to rational approximants of H_ω that possess this same property. Thus we set

$$\hat{\mathcal{R}}_{2n} := \{r \in \mathcal{R}_{2n} : r(z) = r(1/z) \text{ for all } z \in \mathbf{C}\}, \tag{1.5}$$

and put

$$E_{2n,p}(H_\omega) := \inf_{r_{2n} \in \hat{\mathcal{R}}_{2n}} \|H_\omega - r_{2n}\|_{L_p(C)} \tag{1.6}$$

$$= \inf_{r_{2n} \in \hat{\mathcal{R}}_{2n}} \left(\frac{1}{2\pi} \int_{|z|=1} |H_\omega(z) - r_{2n}(z)|^p |dz|\right)^{1/p}. \tag{1.7}$$

For $0 < p < \infty$, we shall show that, for suitable positive constants c_p, C_p,

$$c_p(\sin\omega)^{1/p} n^{1/2p} e^{-\pi\sqrt{n/p}} \leq E_{2n,p}(H_\omega) \leq C_p(\sin\omega)^{1/p} n^{1/2p} e^{-\pi\sqrt{n/p}}. \tag{1.8}$$

4^*

Convergence rates of the form $e^{-c\sqrt{n}}$ also arise in optimal quadrature schemes in the Hardy space H^p (cf. Loeb and Werner [7], Bojanov [3], Stenger [11], Haber [6], Newman [9,10], Andersson [1], and Andersson and Bojanov [2]). Here we set

$$\epsilon_{n,p} := \inf_{\alpha_k,a_k} \sup_{\|f\|_p=1} \left| \int_{-1}^{1} f(x)dx - \sum_{k=1}^{n} a_k f(\alpha_k) \right|, \qquad (1.9)$$

where $a_k, \alpha_k \in \mathbf{C}$, $|\alpha_k| < 1$, and the supremum is taken over all $f \in H^p$ having unit H^p-norm. As explained by Newman in [9], the problem of optimal quadrature is intimately related to best rational approximation. Indeed, we shall show that, for $1 < p < \infty$ and $1/p + 1/q = 1$,

$$c_q E_{2n,q}(H_{\pi/2}) \le \epsilon_{n,p} \le 2\pi E_{2n,q}(H_{\pi/2}) \qquad (1.10)$$

and so (cf. (1.8)) we have

$$c_q n^{1/2q} e^{-\pi\sqrt{n/q}} \le \epsilon_{n,p} \le C_q n^{1/2q} e^{-\pi\sqrt{n/q}}. \qquad (1.11)$$

These inequalities also hold for $p = 1(q = \infty)$ and $p = \infty(q = 1)$, and moreover, the constant C_q in (1.11) can be replaced by an absolute constant. The lower bound in (1.11) (with an absolute constant) was established by Andersson [1], and the upper bound was obtained by Andersson and Bojanov [2] using different methods.

2. Estimates for Approximation of the Signum Function

In this section we establish the asymptotically exact estimates of (1.3). In so doing it is convenient to work instead with rational approximants to the unit step function

$$u(x) := \begin{cases} 1, & 0 < x \le 1 \\ 0, & -1 \le x < 0 \end{cases}. \qquad (2.1)$$

Setting

$$\delta_{n,p}(u) := \inf_{r_n \in \mathcal{R}_n} \|u - r_n\|_{L_p[-1,1]}, \qquad (2.2)$$

we shall prove the following result which is equivalent to (1.3):

THEOREM 2.1. *For $0 < p < \infty$, there exist positive constants c_p, C_p such that*

$$c_p n^{1/2p} e^{-\pi\sqrt{n/p}} \le \delta_{n,p}(u) \le C_p n^{1/2p} e^{-\pi\sqrt{n/p}}, \quad n = 1, 2, \ldots. \qquad (2.3)$$

Throughout this paper, the values of the constants c_p, C_p may vary from one inequality to another.

Proof. As we mentioned in the Introduction, the upper bound in (2.3) was established by Vjacheslavov. Here we only comment (we shall need this later) that Vjacheslavov utilizes approximations of $u(x)$ of the form

$$r(x) = \{1 + P(-x)/P(x)\}^{-1}, \quad P(x) = \prod_{j=0}^{n-1}(x + \xi_j),$$

for suitably chosen points ξ_j ($= \xi_j(p, n)$) in $(0,1)$. The polynomial $P(x)$ satisfies (see [12, formulas (16), (17), (20), (21)])

(a) $|P(-x)/P(x)| \leq 1$ and $|1 + P(-x)/P(x)| \geq 1/2$, $0 \leq x \leq 1$,
(b) $x^{1/p}|P(-x)/P(x)| \leq c_p e^{-\pi\sqrt{np}}$, $0 \leq x \leq 1$.

We turn now to the proof of the lower bound in (2.3). Let $r_n^* \in \mathcal{R}_n$ satisfy

$$\begin{aligned}
\Delta_n := \delta_{n,p}^p(u) &= \int_{-1}^{1} |u(x) - r_n^*(x)|^p dx \\
&= \int_0^1 |r_n^*(-x)|^p dx + \int_0^1 |1 - r_n^*(x)|^p dx.
\end{aligned} \tag{2.4}$$

Since $a^2 + b^2 \geq 2ab$, we have

$$\Delta_n \geq 2 \left(\int_0^1 |r_n^*(-x)|^p dx \right)^{1/2} \left(\int_0^1 |1 - r_n^*(x)|^p dx \right)^{1/2},$$

and so by the Cauchy-Schwarz inequality,

$$\frac{\Delta_n}{2} \geq \int_0^1 |r_n^*(-x)|^{p/2} |1 - r_n^*(x)|^{p/2} dx. \tag{2.5}$$

We write the last integral as

$$\int_0^1 \left| \frac{r_n^*(-x)}{1 - r_n^*(-x)} \cdot \frac{1 - r_n^*(x)}{r_n^*(x)} \cdot r_n^*(x)(1 - r_n^*(-x)) \right|^{p/2} dx$$

and notice that

$$\frac{r_n^*(-x)}{1 - r_n^*(-x)} \cdot \frac{1 - r_n^*(x)}{r_n^*(x)} = \frac{P_{2n}(x)}{P_{2n}(-x)}$$

for some polynomial P_{2n} of degree $\leq 2n$. Then (2.5) becomes

$$\frac{\Delta_n}{2} \geq \int_0^1 \left| \frac{P_{2n}(x)}{P_{2n}(-x)} \right|^{p/2} |r_n^*(x)|^{p/2} |1 - r_n^*(-x)|^{p/2} dx. \tag{2.6}$$

Now let $0 < \epsilon < 1$ (we will make a specific choice for ϵ later) and set

$$A_n = A_n(\epsilon) := \{x \in [\epsilon, 1] : |r_n^*(x)| < 1/2 \text{ or } |r_n^*(-x)| > 1/2\}.$$

From (2.4) we see that $\Delta_n \geq 2^{-p} \cdot \text{meas}(A_n)$; that is, the Lebesgue measure of A_n satisfies

$$\text{meas}(A_n) \leq 2^p \Delta_n. \tag{2.7}$$

Furthermore, from (2.6) and the definition of A_n, we have

$$\frac{\Delta_n}{2} \geq \int_{[\epsilon,1] \setminus A_n} \left| \frac{P_{2n}(x)}{P_{2n}(-x)} \right|^{p/2} |r_n^*(x)|^{p/2} |1 - r_n^*(-x)|^{p/2} dx$$

$$\geq 2^{-p} \int_{[\epsilon,1] \setminus A_n} \left| \frac{P_{2n}(x)}{P_{2n}(-x)} \right|^{p/2} dx \tag{2.8}$$

$$= 2^{-p} \mu_n \int_{[\epsilon,1] \setminus A_n} x \left| \frac{P_{2n}(x)}{P_{2n}(-x)} \right|^{p/2} \frac{dx}{\mu_n x},$$

where μ_n is defined by

$$\mu_n := \int_{[\epsilon,1] \setminus A_n} \frac{dx}{x}. \tag{2.9}$$

Note that from (2.7)

$$\mu_n \geq \log[\epsilon + \text{meas}(A_n)]^{-1} \geq \log[\epsilon + 2^p \Delta_n]^{-1}. \tag{2.10}$$

Observing that $dx/\mu_n x$ is a unit measure, we apply Jensen's inequality to (2.8) and obtain

$$\log(2^{p-1} \Delta_n) \geq \log \mu_n + \int_{[\epsilon,1] \setminus A_n} \log x \frac{dx}{\mu_n x} + \int_{[\epsilon,1] \setminus A_n} \frac{p}{2} \log \left| \frac{P_{2n}(x)}{P_{2n}(-x)} \right| \frac{dx}{\mu_n x}$$

$$\geq \log \mu_n - \frac{1}{2\mu_n} (\log \epsilon)^2 + \frac{p}{2\mu_n} \int_{[\epsilon,1] \setminus A_n} \log \left| \frac{P_{2n}(x)}{P_{2n}(-x)} \right| \frac{dx}{x}. \tag{2.11}$$

To estimate the last integral we use the following inequality due to Newman [8]:

$$\int_a^b \log \left| \frac{x - \xi}{x + \xi} \right| \frac{dx}{x} \geq -\frac{\pi^2}{2}, \tag{2.12}$$

which holds for any $\xi \in \mathbb{C}$ and for all subintervals $[a, b] \subset [-1, 1]$. Writing

$$P_{2n}(x)/P_{2n}(-x) = \prod_{k=1}^{d(n)} (x - \xi_k^{(n)})/(x + \xi_k^{(n)}),$$

where $d(n) := \deg P_{2n}$, we obtain from (2.12)

$$
\int_{[\epsilon,1]\backslash A_n} \log\left|\frac{P_{2n}(x)}{P_{2n}(-x)}\right| \frac{dx}{x} = \sum_{k=1}^{d(n)} \int_{[\epsilon,1]\backslash A_n} \log\left|\frac{x - \xi_k^{(n)}}{x + \xi_k^{(n)}}\right| \frac{dx}{x}
$$

$$
\geq \sum_{k=1}^{d(n)} \int_{[\epsilon,1]\backslash A_n} \log\left|\frac{x - |\mathrm{Re}\,\xi_k^{(n)}|}{x + |\mathrm{Re}\,\xi_k^{(n)}|}\right| \frac{dx}{x} \geq \sum_{k=1}^{d(n)} \int_\epsilon^1 \log\left|\frac{x - |\mathrm{Re}\,\xi_k^{(n)}|}{x + |\mathrm{Re}\,\xi_k^{(n)}|}\right| \frac{dx}{x} \quad (2.13)
$$

$$
\geq -d(n)\frac{\pi^2}{2} \geq -n\pi^2.
$$

Thus, from (2.11) and (2.13), we get

$$
\log(2^{p-1}\Delta_n) \geq \log\mu_n - \frac{1}{2\mu_n}(\log\epsilon)^2 - \frac{pn\pi^2}{2\mu_n}. \quad (2.14)
$$

To obtain the desired lower bound, suppose to the contrary that

$$
\Delta_n = \delta_{n,p}^p(u) = \tau_n n^{1/2} e^{-\pi\sqrt{pn}}, \quad (2.15)
$$

where $\tau_n \to 0$ as $n \to \infty$. Choose $\epsilon = n^{1/2} e^{-\pi\sqrt{pn}}$. Then (cf. (2.10))

$$
\mu_n \geq \log[\epsilon + 2^p\Delta_n]^{-1} = \log\epsilon^{-1} + \log[1 + 2^p\tau_n]^{-1}
$$

$$
= \pi\sqrt{pn} - \frac{1}{2}\log n - \log(1 + 2^p\tau_n), \quad (2.16)
$$

which implies that

$$
\log\mu_n \geq \frac{1}{2}\log n + O(1). \quad (2.17)
$$

Also, by (2.16),

$$
-\frac{1}{2\mu_n}(\log\epsilon)^2 = -\frac{1}{2}\log\epsilon^{-1} \cdot \frac{\log\epsilon^{-1}}{\mu_n} \geq -\frac{1}{2}\log\epsilon^{-1} \cdot \frac{\log\epsilon^{-1}}{\log\epsilon^{-1} + \log[1 + 2^p\tau_n]^{-1}}
$$

$$
= -\frac{1}{2}\log\epsilon^{-1} + o(1),
$$

that is,

$$
-\frac{1}{2\mu_n}(\log\epsilon)^2 \geq \frac{1}{4}\log n - \frac{\pi}{2}\sqrt{pn} + o(1). \quad (2.18)
$$

Furthermore, by (2.16),

$$-\frac{pn\pi^2}{2\mu_n} = -\frac{\pi}{2}\sqrt{pn} \cdot \frac{1}{\mu_n/(\pi\sqrt{pn})}$$

$$\geq -\frac{\pi}{2}\sqrt{pn} \cdot \frac{1}{[\pi\sqrt{pn} - \frac{1}{2}\log n - \log(1 + 2^p\tau_n)]/(\pi\sqrt{pn})} \qquad (2.19)$$

$$= -\frac{\pi}{2}.\sqrt{pn}\left\{1 + \frac{1}{2}\frac{\log n}{\pi\sqrt{pn}}\right\} + o(1).$$

Finally, substituting into (2.14) the expression for Δ_n in (2.15) and utilizing the estimates of (2.17), (2.18) and (2.19) we get

$$\log\tau_n \geq O(1),$$

which yields the desired contradiction. ∎

3. Estimates for $E_{2n,p}(H_\omega)$

With $H_\omega, \hat{\mathcal{R}}_{2n}$ and $E_{2n,p}(H_\omega)$ as defined in (1.4), (1.5) and (1.6), respectively, we first establish

LEMMA 3.1. For $0 < \omega < \pi$, $0 < p < \infty$, $n = 1, 2, \dots$, there holds

$$E^p_{2n,p}(H_\omega) = \inf_{r_n \in \mathcal{R}_n} \frac{1}{\pi}\int_{-1}^1 |u(x) - r_n(x)|^p \frac{\sin\omega}{1 + x\cos\omega}\frac{dx}{\sqrt{1-x^2}}, \qquad (3.1)$$

where u is defined in (2.1).

Proof. Let $\hat{r}_{2n} \in \hat{\mathcal{R}}_{2n}$ satisfy

$$E^p_{2n,p}(H_\omega) = \frac{1}{2\pi}\int_{|z|=1} |H_\omega(z) - \hat{r}_{2n}(z)|^p|dz|.$$

The bilinear transformation

$$z = \frac{i\lambda - t}{i\lambda + t}, \quad \lambda := \cot(\omega/2), \qquad (3.2)$$

maps $C : |z| = 1$ onto $(-\infty, \infty)$ and transforms the arc $\{z = e^{i\theta} : |\theta| < \omega\}$ to the line segment $(-1, 1)$. Thus,

$$E^p_{2n,p}(H_\omega) = \frac{1}{2\pi}\int_{-\infty}^{\infty} |U(t) - R_{2n}(t)|^p \frac{2\lambda}{\lambda^2 + t^2}dt, \qquad (3.3)$$

where $R_{2n}(t) := \hat{r}_{2n}(z)$ and

$$U(t) := \begin{cases} 1, & \text{if } -1 < t < 1 \\ 0, & \text{if } |t| > 1 \end{cases}$$

We note that since $\hat{r}_{2n}(z) = \hat{r}_{2n}(1/z)$, the rational function $R_{2n} \in \mathcal{R}_{2n}$ is even, i.e. $R_{2n}(t) = r_n(t^2)$ for some $r_n \in \mathcal{R}_n$. Thus (3.3) can be written as

$$E_{2n,p}^p(H_\omega) = \frac{1}{2\pi} \int_{-\infty}^{\infty} |U(t) - r_n(t^2)|^p \frac{2\lambda}{\lambda^2 + t^2} dt$$

$$= \frac{1}{\pi} \int_0^{\infty} |U(t^2) - r_n(t^2)|^p \frac{2\lambda}{\lambda^2 + t^2} dt.$$

On setting $x = (1 - t^2)/(1 + t^2)$ in the last integral and using the definition of λ in (3.2), we then get

$$E_{2n,p}^p(H_\omega) = \frac{1}{\pi} \int_{-1}^1 \left| u(x) - r_n\left(\frac{1-x}{1+x}\right) \right|^p \frac{\sin \omega}{1 + x \cos \omega} \frac{dx}{\sqrt{1 - x^2}},$$

which proves that $E_{2n,p}^p(H_\omega)$ is at least as large as the right-hand side of (3.1). The reverse inequality follows in a similar way. ∎

We are now in position to establish

THEOREM 3.2. For $0 < \omega < \pi$, $0 < p < \infty$ and $n = 1, 2, \ldots$,

$$c_p(\sin \omega)^{1/p} n^{1/2p} e^{-\pi\sqrt{n/p}} \leq E_{2n,p}(H_\omega) \leq C_p(\sin \omega)^{1/p} n^{1/2p} e^{-\pi\sqrt{n/p}}, \quad (3.4)$$

where the positive constants c_p, C_p are independent of ω and n.

Proof. The lower estimate is straightforward. Let $r_n^* \in \mathcal{R}_n$ satisfy

$$E_{2n,p}^p(H_\omega) = \frac{1}{\pi} \int_{-1}^1 |u(x) - r_n^*(x)|^p \frac{\sin \omega}{1 + x \cos \omega} \frac{dx}{\sqrt{1 - x^2}}.$$

Since $(1 + x \cos \omega)^{-1} \geq 1/2$ for $x \in [-1, 1]$, we get from Theorem 2.1

$$E_{2n,p}^p(H_\omega) \geq \frac{\sin \omega}{2\pi} \int_{-1}^1 |u(x) - r_n^*(x)|^p \, dx$$

$$\geq \frac{\sin \omega}{2\pi} \delta_{n,p}^p(u) \geq c_p(\sin \omega)\sqrt{n} e^{-\pi\sqrt{np}}.$$

To establish the upper estimate, care must be taken when ω is near 0 or π. For this purpose we choose an integer k so large that $kp > 1$. Let $P(x)$ be the polynomial of degree n defined in the beginning of Section 2 and set

$$R_{n+k}(x) := \left\{ 1 + \frac{P(-x)}{P(x)} \left(\frac{1-x}{1+x}\right)^k \right\}^{-1} \in \mathcal{R}_{n+k}. \quad (3.5)$$

Since $|(1-x)/(1+x)| \leq 1$ on $[0,1]$, the estimates (a) for $P(x)$ imply that

$$|1 - R_{n+k}(x)| \leq 2|P(-x)/P(x)|, \quad 0 \leq x \leq 1.$$

Applying estimate (b) we then obtain (cf. [12, p. 30]):

$$\int_{-1}^{1} |u(x) - R_{n+k}(x)|^p dx = 2 \int_{0}^{1} |1 - R_{n+k}(x)|^p dx \leq C_p \sqrt{n} e^{-\pi\sqrt{np}}. \quad (3.6)$$

By Lemma 3.1,

$$E_{2n+2k,p}^{p}(H_\omega) \leq \frac{1}{\pi} \int_{-1}^{1} |u(x) - R_{n+k}(x)|^p \frac{\sin\omega}{1 + x\cos\omega} \frac{dx}{\sqrt{1-x^2}}. \quad (3.7)$$

To estimate the last integral, we first assume $|x| \leq 1/2$. Then $(1+x\cos\omega)^{-1} \leq 2$ and $1/\sqrt{1-x^2} \leq 2/\sqrt{3}$ so that by (3.6)

$$\frac{1}{\pi} \int_{-1/2}^{1/2} |u(x) - R_{n+k}(x)|^p \frac{\sin\omega}{1 + x\cos\omega} \frac{dx}{\sqrt{1-x^2}} \leq \frac{4}{\pi\sqrt{3}} C_p(\sin\omega)\sqrt{n} e^{-\pi\sqrt{np}}.$$

$$(3.8)$$

For $1/2 \leq x \leq 1$, we again use the estimates (a), (b) to obtain

$$|u(x) - R_{n+k}(x)|^p = \left| \frac{P(-x)}{P(x)} \right|^p \left| \frac{1-x}{1+x} \right|^{kp} \left| 1 + fracP(-x)P(x)\left(\frac{1-x}{1+x}\right)^k \right|^{-p}$$

$$\leq c_p |1-x|^{kp} e^{-\pi\sqrt{np}},$$

Since $kp > 1$, we have $|1-x|^{kp}/(1+x\cos\omega) \leq 1$ on $[1/2, 1]$ and thus

$$\frac{1}{\pi} \int_{1/2}^{1} |u(x) - R_{n+k}(x)|^p \frac{\sin\omega}{1 + x\cos\omega} \frac{dx}{\sqrt{1-x^2}}$$

$$(3.9)$$

$$\leq c_p(\sin\omega) e^{-\pi\sqrt{np}} \int_{1/2}^{1} \frac{|1-x|^{kp}}{1+x\cos\omega} \frac{dx}{\sqrt{1-x^2}}$$

$$\leq C_p(\sin\omega) e^{-\pi\sqrt{np}}.$$

Furthermore, the same estimate holds for integration over $[-1, -1/2]$. Hence we get from (3.7), (3.8) and (3.9)

$$E_{2n+2k,p}^{p}(H_\omega) \leq C_p(\sin\omega)\sqrt{n} e^{-\pi\sqrt{np}},$$

which completes the proof. ∎

Remark. For $p \geq 1$ one can replace the constant C_p in the upper bound of (3.4) by an absolute constant as follows from Vjacheslavov's results (see the Introduction).

4. Optimal Quadrature in H^p

For functions in the Hardy space H^p, $1 \leq p \leq \infty$, of the unit disk $\Delta : |z| < 1$, the error in an optimal n-term quadrature formula is defined by

$$\epsilon_{n,p} := \inf_{\alpha_k, a_k} \sup_{\|f\|_p = 1} \left| \int_{-1}^{1} f(x)dx - \sum_{k=1}^{n} a_k f(\alpha_k) \right|, \tag{4.1}$$

where the infimum is taken over all $a_k, \alpha_k \in \mathbf{C}$ with $|\alpha_k| < 1$, $k = 1, 2, \ldots, n$, and the supremum is over the unit ball in H^p. In (4.1) we allow that a point α_j be repeated; in such a case, the summation in (4.1) is to be modified by including the appropriate number of successive derivatives of f evaluated at α_j.

THEOREM 4.1. *Let $1 < p < \infty$ and $1/q + 1/p = 1$. Then there exists a positive constant c_q such that*

$$c_q E_{2n,q}(H_{\pi/2}) \leq \epsilon_{n,p} \leq 2\pi E_{2n,q}(H_{\pi/2}), \quad n = 1, 2, \ldots, \tag{4.2}$$

where $E_{2n,q}$ is defined in (1.6).

Proof. Let $|\alpha_k| < 1$, $k = 1, \ldots, n$. As observed by T. Rivlin (cf. [9, p.13]), we can use the Cauchy integral formula to write

$$\int_{-1}^{1} f(x)dx - \sum_{k=1}^{n} a_k f(\alpha_k) = \int_{|z|=1} f(z) \left(H(z) - \frac{1}{2\pi i} \sum_{k=1}^{n} \frac{a_k}{z - \alpha_k} \right) dz, \tag{4.3}$$

where $H(z)$ is the Heaviside function, defined to be 0 on the upper unit semi-circle and 1 on the lower semi-circle. By the duality relationship (cf. [D, p.132]), we have

$$\sup_{\|f\|_p = 1} \left| \int_{|z|=1} f(z)(H(z) - R_n(z))dz \right| = 2\pi \inf_{g \in H^q} \|H - R_n - g\|_q, \tag{4.4}$$

where

$$R_n(z) := \frac{1}{2\pi i} \sum_{k=1}^{n} \frac{a_k}{z - \alpha_k}. \tag{4.5}$$

(For the case of repeated α_k's, we modify the form of R_n accordingly so as to have multiple poles.) From (4.1), (4.3) and (4.4) we get, after rotation by $\pi/2$,

$$\epsilon_{n,p} = 2\pi \inf_{R_n} \inf_{g \in H^q} \|H_{\pi/2}(z) - R_n(z/i) - g(z)\|_q, \tag{4.6}$$

where the first infimum is taken over all R_n of the form (4.5).

To obtain the upper estimate, we take $g(z) = a_0 + R_n(1/(iz)) \in H^q$, so that from (4.6),

$$\epsilon_{n,p} \leq 2\pi \inf_{R_n, a_0} \|H_{\pi/2}(z) - a_0 - R_n(z/i) - R_n(1/(iz))\|_q. \qquad (4.7)$$

Since $a_0 + R_n(z/i) + R_n(1/(iz)) \in \hat{R}_{2n}$ (see (1.5)) and since an arbitrary $r_{2n} \in \hat{R}_{2n}$ analytic on $|z| = 1$ can be written in such a form for suitable a_k's and α_k's, we see that the right-hand side of (4.7) equals $2\pi E_{2n,q}(H_{\pi/2})$. Thus

$$\epsilon_{n,p} \leq 2\pi E_{2n,q}(H_{\pi/2}). \qquad (4.8)$$

To obtain the lower estimate in (4.2), consider the Fourier representation

$$H_{\pi/2}(z) = \left(b_0 + \sum_{k=1}^{\infty} b_k z^k\right) + \sum_{k=1}^{\infty} b_k z^{-k} =: H_{\pi/2}^+(z) + H_{\pi/2}^-(z),$$

where we have used the property that $H_{\pi/2}(1/z) = H_{\pi/2}(z)$ for $|z| = 1$. For R_n of the form (4.5) (so that R_n is analytic outside $C : |z| = 1$), we obtain as a consequence of the M. Riesz theorem that

$$\|H_{\pi/2}(z) - R_n(z/i) - g(z)\|_q \geq c_q \|H_{\pi/2}^-(z) - R_n(z/i)\|_q, \qquad (4.9)$$

where $g \in H^q$ and $c_q > 0$ depends only on q. Since

$$H_{\pi/2}^+(z) = b_0 + H_{\pi/2}^-(1/z), \quad |z| = 1,$$

we have

$$\|H_{\pi/2}^-(z) - R_n(z/i)\|_q = \|H_{\pi/2}^+(z) - b_0 - R_n(1/(iz))\|_q. \qquad (4.10)$$

Thus, from (4.9), (4.10) and the triangle inequality, we get

$$\|H_{\pi/2}(z) - R_n(z/i) - g(z)\|_q \geq \frac{c_q}{2}\|H_{\pi/2}(z) - b_0 - R_n(z/i) - R_n(1/(iz))\|_q$$

$$\geq \frac{c_q}{2} E_{2n,q}(H_{\pi/2}),$$

and so the desired lower bound follows from (4.6). ∎

Remark. Since the duality relation (4.4) holds also for $p = \infty$ ($q = 1$), the estimate (4.8) is valid in this case, i.e.,

$$\epsilon_{n,\infty} \leq 2\pi E_{2n,1}(H_{\pi/2}), \quad n = 1, 2, \ldots . \qquad (4.11)$$

As a consequence of Theorems 3.2 and 4.1 and the inequality (4.11) we obtain the following result due to Andersson and Bojanov [2].

THEOREM 4.2. Let $1 \leq p \leq \infty$ and $1/q + 1/p = 1$. Then there exist positive constants c_q, C_0 such that

$$c_q n^{1/2q} e^{-\pi\sqrt{n/q}} \leq \epsilon_{n,p} \leq C_0 n^{1/2q} e^{-\pi\sqrt{n/q}}, \quad n = 1, 2, \ldots . \qquad (4.12)$$

We remark that the lower bound in (4.12) *with an absolute constant* c_0 follows from Newman's methods [9] as shown by Andersson [1].

References

[1] J.-E. Andersson, *Optimal quadrature of H^p functions*, Math Z. **172** (1980), 55–62.

[2] J.-E. Andersson and B.D. Bojanov, *A note on the optimal quadrature in H^p*, Numer.Math. **44** (1984), pp. 301–308.

[3] B. Bojanov, *On an optimal quadrature formula*, C.R. Acad. Bulgare Sci. **27** (1974), pp. 619–621.

[4] A.A. Gonchar, *On the rapidity of rational approximation of continuous functions with characteristic singularities*, Mat. Sb **73** (1967), pp. 630–638; Math. USSR. Sb. **2** (1967), pp. 561–568.

[5] A.A. Gonchar, *Estimates of the growth of rational functions and some of their applications*, Mat. Sb **72** (1967); Math. USSR. Sb. **1** (1967), pp. 445–456.

[6] S. Haber, *The tanh rule for numerical integration*, SIAM J. Numer. Anal. **14** (1977), pp. 668–685.

[7] H.L. Loeb and H. Werner, *Optimal numerical quadrature in H^p spaces*, Math. Z. **138** (1974), pp. 111–117..

[8] D.J. Newman, *Rational approximation to $|x|$*, Michigan Math. J. **111** (1964), pp. 11–14..

[9] D.J. Newman, *Approximation with Rational Functions*, Regional Conf. Series in Math. **41** (1979.), Amer. Math. Soc., Providence.

[10] D.J. Newman, *Quadrature formulae for H^p functions*, Math. Z. **166** (1979), pp. 111–115..

[11] F. Stenger, *Optimal convergence of minimum norm approximations in H^p*, Numer. Math. **29** (1978), pp. 345–362..

[12] N.S. Vjacheslavov, *On the least deviations of the function* sgn x *and its primitives from the rational functions in the L_p metrics*, $0 < p < \infty$, Mat. Sb. **103** (1977); Math. USSR Sb. **32** (1977), pp. 19–31.

Department of Mathematics
The Open University of Israel
P.O. Box 39328
Tel Aviv, ISRAEL

[2]Institute for Constructive Mathematics
Department of Mathematics
University of South Florida
Tampa, FL 33620, USA

UNIFORM RATIONAL APPROXIMATION OF $|X|$

Herbert Stahl*

We review recent results about best uniform rational approximants of $|x|$ on $[-1, 1]$. We shall sketch the proof of the theorem that

$$\lim_{n \to \infty} e^{\pi \sqrt{n}} E_{nn}(|x|, [-1, 1]) = 8,$$

where $E_{nn}(|x|, [-1, 1])$ denotes the minimal approximation error. Related results are concerned with the asymptotic distribution of poles and zeros of the approximants r_n^* and the distribution of extreme points of the error function $|x| - r_n^*(x)$ on $[-1, 1]$. Two conjectures about a generalisation of the approximation problem of $|x|$ on $[-1, 1]$ or of x^α, $\alpha > 0$, on $[0, 1]$ will be formulated.

1. Introduction

The development of approximation theory has been strongly influenced by the study of special functions. Some of these functions have attracted an impressive amount of attention. Despite of their special character many of these investigations have given valuable insight into the nature of the approximation problem, and new techniques have been introduced in connection with such studies. One of the prominent special functions in approximation theory is $|x|$ and its approximation by polynomials or rational functions on the intervall $[-1, 1]$. Before we come to the discussion of new results we will review some of the earlier contributions.

Let Π_n denote the set of all polynomials of degree at most $n \in \mathbb{N}$ with real coefficients, let $\mathcal{R}_{mn}$ denote the set $\{p/q \mid p \in \Pi_m, \ q \in \Pi_n, \ q \not\equiv 0\}$, $m, n \in \mathbb{N}$, of rational functions, and let the best rational approximant $r_n^* \in \mathcal{R}_{nn}$ and the approximation error $E_{nn} = E_{nn}(|x|, [-1, 1])$ be defined by

$$E_{nn} = \||x| - r_n^*\|_{[-1,1]} = \inf_{r \in \mathcal{R}_{nn}} \||x| - r\|_{[-1,1]}, \tag{1.1}$$

*Research supported by the Deutsche Forschungsgemeinschaft (AZ: Sta 299 14–1)
AMS (MOS) *subject classification: 41A20,25,50.*
Key words: *Best rational approximation, distribution of poles, zeros, and extreme points.*

where $\| \cdot \|_K$ denotes the sup-norm on $K \subseteq \mathbb{R}$. It is well-known that the best approximant r_n^* exists and is unique (see [12], §9.1, or [14], §5.1). We shall here be concerned mainly with the speed of convergence of the sequence $\{E_{nn}\}_{n \in \mathbb{N}}$. In order to understand better the convergence behavior we will also investigate the asymptotic distribution of poles and zeros of the approximats r_n^* and the distribution of extreme points of the error function $|x| - r_n^*(x)$ on $[-1, 1]$. Most of the proofs will only be sketched, while complete proofs can be found in [16], where the convergence result has been proved, and in [17], where the distributional results have been proved. We begin with some histotric remarks.

S. Bernstein has devoted several research papers to the study of polynomial approximation and interpolation of the function

$$f(x) = |x| \qquad (1.2)$$

on the interval $[-1, 1]$. Most interesting for our discussion is his classical paper from 1913 in [2], in which he has shown that the limit

$$\lim_{n \to \infty} n E_n(|x|, [-1, 1]) = \beta \qquad (1.3)$$

exists. In (1.3) $E_n = E_n(|x|, [-1, 1])$ denotes the minimal error with respect to polynomial approximation. He gave the numerical bounds $0.278 < \beta < 0.286$ for β, and made, based on these bounds, the conjecture that β may be equal to the number $1/(2\sqrt{\pi}) = 0.282094\ldots$ However, this conjecture has meanwhile been disproved. In [19] an extensive numerical investigation of the convergence of E_n, $n \to \infty$, has been carried out and the value of β has been calculated up to fifty significant digits. This numerical investigation disproved Bernstein's conjecture. So far only a numerical value is known for the constant β.

The investigation of rational approximants to f was started by D.J. Newman [13] in 1964 with the surprising (at the time) result that

$$\tfrac{1}{2} e^{-9\sqrt{n}} \leq E_{nn}(|x|, [-1, 1]) \leq 3 e^{-\sqrt{n}} \qquad \text{for} \quad n = 4, 5, \ldots \qquad (1.4)$$

A comparison of the upper estimates in (1.4) and (1.3) show that the speed of rational approximation is incomparable faster than that of polynomial approximation.

Newmann's result has triggered a series of contributions in the last 25 years. Besides of the rational approximation of $|x|$ on $[-1, 1]$ they were also concerned with the closely related problem of approximating x^α, $\alpha > 0$, on $[0, 1]$. We have taken the following compilation from [22]:

$$E_{nn}(x^\alpha, [0, 1]) \leq e^{-c(\alpha)\sqrt[3]{n}}, \alpha \in \mathbb{R}_+, \text{ in } [6], 1967,$$

$$E_{nn}(x^{1/3}, [0, 1]) \leq e^{-c\sqrt{n}}, \text{ in } [4], 1968,$$

$$E_{nn}(x^\alpha, [0, 1]) \leq e^{-c(\alpha)\sqrt{n}}, \alpha \in \mathbb{R}_+, \text{ in } [8], 1967,$$

$$\tfrac{1}{3} e^{-\pi\sqrt{2n}} \leq E_{nn}(x^{1/2}, [0, 1]) \leq e^{-\pi\sqrt{2n}(1 - O(n^{-1/4}))}, \text{ in } [5], 1968,$$

$$e^{-c(\alpha)\sqrt{n}} \le E_{nn}(x^\alpha, [0,1]), \alpha \in Q_+ \setminus \mathbb{N}, \text{ in } [9], 1972,$$

$$e^{-4\pi\sqrt{\alpha n}(1+\epsilon)} \le E_{nn}(x^\alpha, [0,1]) \le e^{-\pi\sqrt{\alpha n}(1-\epsilon)}, \ \alpha \in \mathbb{R}_+ \setminus \mathbb{N}, \epsilon > 0,$$

$$n \ge n(\epsilon, \alpha), \text{ in } [10], 1974,$$

$$E_{nn}(x^{1/2}, [0,1]) \le cne^{-\pi\sqrt{2n}}, \text{ in } [20], 1974,$$

$$1/3 e^{-\pi\sqrt{2n}} \le E_{nn}(x^{1/2}, [0,1]) \le ce^{-\pi\sqrt{2n}}, \text{ in } [21], 1975,$$

$$e^{-c_1(S)\sqrt{n}} \le E_{nn}(\sqrt[s]{n}, [0,1]) \le e^{-c_2(S)\sqrt{n}}, \ s \in \mathbb{N}, \text{ in } [18], 1976.$$

Here, c, $c(\alpha)$, ... denote constants. The best result with respect to the approximation of x^α on $[0,1]$ has been obtained independently by T. Ganelius in [7] in 1979 and by N.S. Vjacheslavov in [22] in 1980; they have proved that

$$c_1(\alpha)e^{-2\pi\sqrt{\alpha n}} \le E_{nn}(x^\alpha, [0,1]) \le c_2(\alpha)e^{-2\pi\sqrt{\alpha n}} \text{ for } n = 1, 2, \ldots \quad (1.5)$$

However, the upper bound in (1.5) has only been proved for $\alpha \in Q_+$, and it could not be shown that $c_2(\alpha)$ depends continuously on α. The lower bound in (1.5) holds for all $\alpha \in \mathbb{R}_+ \setminus \mathbb{N}$.

In [22] and [1] also approximation in L^p-norms has been investigated, both papers contain very sharp results in this respect, but we will here not discuss results in L^p-norms. There is an other result in [1], which we will mention since it is of special interest with respect to the conjecture that will be formulated in §4 below: The lower bound of (1.5) has been extended to Markow functions of type x^α, $\alpha > 0$, i.e., functions of the form

$$f(z) = \int_0^\infty \frac{\varphi(x)dx}{z+x}, \quad (1.6)$$

where φ is a positive function satisfying

$$0 < c_1 \le x^{-\alpha}\varphi(x) \le c_2 < \infty \quad \text{for } x \in [0,\epsilon], \quad \epsilon > 0. \quad (1.7)$$

It is rather immediate that $-z^\alpha$, $0 < \alpha < 1$, is of type (1.6).

From an historical point of view it may be interesting to repeat an observation in [9], where it was pointed out that Newman's result can be deduced rather immediately from an old result (obtained in 1877) by Zolotarev about best rational approximation of the function $\text{sign}(x)$ on the union of the two intervals $[-1, -h] \cup [h, 1]$, $0 < h < 1$ (Zolotarev's fourth problem). Thus, in some sense the problem of rational approximation of $|x|$ on $[-1, 1]$ has a history that dates back even earlier than Bernstein's work on polynomial approximation of $|x|$. From Zolotarev's result it follows (see [9] for details) that

$$e^{-\pi(\sqrt{2}+\epsilon)\sqrt{n}} \le E_{nn}(|x|, [-1,1]) \le e^{-\pi(\frac{1}{\sqrt{2}}-\epsilon)\sqrt{n}} \quad (1.8)$$

for $\epsilon > 0$ and $n \geq n_0(\epsilon)$. This is a sharpening of (1.4) and, moreover, the result hints already towards the correct coefficient $-\pi$ in the exponent of the error estimate.

The strongest result with respect to rational approximation of $|x|$ on $[-1,1]$, which was known until recently, has been proved by N.S. Vjacheslavov in [22] using and extending results from [5] and [21]. The result is a special case of that mentioned in (1.5), and it says that there exist constants M_1, M_2 such that

$$0 < M_1 \leq e^{\pi\sqrt{n}} E_{nn}(|x|, [-1,1]) \leq M_2 < \infty \qquad (1.9)$$

for $n = 2, 3, \ldots$ From [5] one can deduce that $M_1 \geq 1/3$.

The inequalities (1.9) prove that $-\pi$ is the exact coefficient in the exponent of the error estimate. However, in [22] the question remains open whether a strong error formula exits that is the rational analogue to Bernstein's result (1.3).

Recently R.S. Varga, A. Ruttan, and A.J. Carpenter [23] have numerically investigated the errors $E_{nn}(|x|, [-1,1])$ for $n \to \infty$. Following a similar path as in [19] they have calculated the numbers $E_{nn} = E_{nn}(|x|, [-1,1])$ for even n up to $n = 80$ with high precision and then applied a Richardson extrapolation to the sequence $\{E_{2n,2n}\}_{n=1}^{10}$. The extrapolation is based on the assumption that an expansion of the form

$$e^{\pi\sqrt{2n}} E_{2n,2n}(|x|, [-1,1]) \approx K_0 + \frac{K_1}{\sqrt{n}} + \frac{K_2}{n} + \frac{K_3}{n^{3/2}} + \ldots \qquad (1.10)$$

is valid. The analysis results in numerical values for the constants K_0, $K_1, \ldots, K_4$, and it gives strong evidence that K_0 has the exact value 8. Hence, it was natural to conjecture that the limit

$$\lim_{n\to\infty} e^{\pi\sqrt{n}} E_{nn}(|x|, [-1,1]) \qquad (1.11)$$

exists and is equal 8. This conjecture has been proved recently by the author, and the full proof can be found in [16]. In the next section we shall discuss the basic ideas of the proof. After that in section 3 we shall investigate the asymptotic distribution of poles and zeros of the approximants r_n^* and the distribution of extreme points of the error function $|x| - r_n^*(x)$ on $[-1,1]$. In section 4 we discuss two conjectures about a natural generalizations of the results presented here.

2. A Strong Error Estimate

In the present section we shall discuss the proof of

THEOREM 1. ([16], Theorem 1) *We have*

$$\lim_{n\to\infty} e^{\pi\sqrt{n}} E_{nn}(|x|, [-1,1]) = 8. \qquad (2.1)$$

The limit (2.1) is a direct analogue of Bernstein's result (1.3) for rational approximation. Somewhat astonishing is the fact that here the constant 8 is a natural number, while in the presumably more elementary case of polynomial approximation the nature of the constant β is still not understood, but certainly different from a natural number.

The proof of Theorem 1 can be divided in two parts. In the first one we study three consecutive transformations in which the error function $|x|-r_n^*(x)$ is brought in a form that can be compared with a certain Green potential. The comparison then constitutes the second part of the proof. The properties of the Green potential will be stated in Theorem 2 below.

The first transformation is given by

$$r_n(x) := \frac{r_n^*(x) - x}{r_n^*(x) + x} \in \mathcal{R}_{n+1,n+1}. \tag{2.2}$$

We assume that n is even. It can be shown by standard methods (basically Chebyshev's Theorem on alternants is used) that the error function $|x| - r_n^*$ has exactly $n+2$ extreme points $\{x_j\}_{j=0}^{n+1}$ in $[0,1]$, i.e. there are $n+1$ points

$$0 = x_0 < x_1 < \cdots < x_{n+1} = 1 \tag{2.3}$$

with the property that

$$r_n^*(x_j) - x_j = (-1)^j \epsilon_n, \quad j = 0, \ldots, n+1, \tag{2.4}$$

where ϵ_n is used as an abbreviation for the minimal error $E_{nn}(|x|, [-1,1])$. From (2.4) it immediately follows that all $n+1$ zeros of r_n have to be simple and are contained in $(0,1)$. From (2.2) it then follows that the $n+1$ poles of r_n are the mirror images of the zeros with respect to reflection on $i\mathbb{R}$ and lie all on the interval $(-1,0)$. For the extreme points $\{x_j\}$ we have

$$r_n(x_j) = \begin{cases} \dfrac{\epsilon_n}{2x_j + \epsilon_n} & \text{for } j = 0, 2, \ldots, n \\[2mm] \dfrac{-\epsilon_n}{2x_j - \epsilon_n} & \text{for } j = 1, 3, \ldots, n+1. \end{cases} \tag{2.5}$$

On the imaginary axis we have

$$|r_n(iy)| = 1 \qquad \text{for all } y \in \mathbb{R}. \tag{2.6}$$

The location of poles and zeros of r_n shows that $|r_n|$ is small on $(0,1]$ and large on $[-1,0)$. However, (2.5) also shows that the smallness and largeness does not hold uniformly on $(0,1]$ and $[-1,0)$, respectively. The next transformation has been introduced in order to establish such an uniformity. The new rational function R_n is defined as

$$R_n(z) := \frac{4z^2 - 1}{z} r_n(\epsilon_n z) + \frac{1}{z}. \qquad (2.7)$$

In the transition from r_n to R_n the extreme points x_j, $j = 1, \ldots, n+1$, are transformed into

$$z_j := \frac{x_j}{\epsilon_n}, \quad j = 1, \ldots, n+1. \qquad (2.8)$$

Implicite in (2.7) a transformation $x \mapsto z := x/\epsilon_n$ of the independent variable has been defined. Under this transformation the interval $[0,1]$ is mapped onto the intervall $[0, 1/\epsilon_n]$. In the z–variable the point $1/2$ plays a special role, we have

$$R_n(1/2) = 2 \quad \text{and} \quad R_n(z) > 2 \quad \text{for } z \in (0, 1/2). \qquad (2.9)$$

The function R_n assumes its extreme values on $[1/2, 1/\epsilon_n]$ at points z_j, $j = 1, \ldots, n+1$. These points have been defined in (2.8). We have

$$R_n(z_j) = (-1)^j 2, \qquad j = 1, \ldots, n+1, \qquad (2.10)$$

and

$$R_n(z) \in [-2, 2] \qquad \text{for all} \quad z \in [1/2, 1/\epsilon_n] \qquad (2.11)$$

(For more details and full proofs see [16], Lemma 2.3).

From (2.10) we see that the function R_n has extreme values on the interval $[1/2, 1/\epsilon_n]$ of equal size. This property was the main purpose for the specific form of the transformation (2.7). In order to have a function of constant modulus on the interval $[1/2, 1/\epsilon_n]$ we define

$$\Phi_n(z) := \frac{1}{2} \left(R_n(z) + \sqrt{R_n(z)^2 - 4} \right) \qquad \text{for} \quad z \in H_+ \setminus [1/2, 1/\epsilon_n], \quad (2.12)$$

where $H_+ := \{z \mid \text{Re}(z) > 0\}$, and the branch of the root in (2.12) is chosen in such a way that the root is positive for $z \in (0, 1/2)$. The function Φ_n is analytic in $H_+ \setminus [1/2, 1/\epsilon_n]$ and has continuous boundary values on $\partial H_+ \cup [1/2, 1/\epsilon_n]$. On $[1/2, 1/\epsilon]$ we have different boundary values for continuation from both sides, but their modulus are equal. It follows from (2.12) and properties of R_n stated in (2.10) and (2.11) (see also [16], Lemma 2.4) that

$$|\Phi_n(z)| = 1 \quad \text{for all} \quad z \in [\frac{1}{2}, \frac{1}{\epsilon_n}] \qquad (2.13)$$

and

$$
|\Phi_n(iy)| \begin{cases} \leq 4(|y| + |y|^{-1}) & \text{for } y \in \mathbb{R} \\ \geq 2|y| \left(1 + \sqrt{1 - \frac{1}{4}|y|^{-2}}\right) & \text{for } y \in \mathbb{R} \setminus (-\frac{1}{2}, \frac{1}{2}) \\ \geq 2|y| & \text{for } y \in (-\frac{1}{2}, \frac{1}{2}). \end{cases} \tag{2.14}
$$

There exists a positive measure μ_n with

$$
\text{supp}(\mu_n) = \left[\frac{1}{2}, \frac{1}{\epsilon_n}\right], \quad \|\mu_n\| = n + 1, \tag{2.15}
$$

and a function φ_n harmonic in H_+ with

$$
\varphi_n(iy) = \log |\Phi_n(iy)| \quad \text{for } y \in \mathbb{R} \tag{2.16}
$$

so that

$$
\log |\Phi_n(z)| = \varphi_n(z) + \int \log \left|\frac{z - x}{z + x}\right| d\mu_n(x) \quad \text{for } z \in H_+. \tag{2.17}
$$

We note that the bounds in (2.14) are independent of n, and they are sharp near infinity. We have

$$
|\Phi_n(z)| \approx 4|z| \quad \text{for } z \text{ near } \infty. \tag{2.18}
$$

Representation (2.17) of $\log |\Phi_n(z)|$ will be of great importance in the sequel.

With the definition and the short study of the function Φ_n the first part of the proof is completed. We would like to remark that the ideas leading to the definition of Φ_n are basically the same as those leading to the definition of the Chebyshev polynomials by

$$
T_n(x) := \zeta^n + \zeta^{-n}, \quad \zeta = \frac{1}{2}(x + \sqrt{x^2 - 4}), \quad x \in [-2, 2]. \tag{2.19}
$$

Here also $|\zeta| = $ constant for $x \in [-2, 2]$, and therefore identity (2.19) rather easily allows to obtain asymptotics for T_n as $n \to \infty$. Thus, Φ_n plays the same role as ζ, only the connection with the original approximant r_n^* is more complicated

The function Φ_n, and especially its representation (2.17), allows us to compare the function

$$
\log \left|\frac{1}{4z}\Phi_n(z)\right| \tag{2.20}
$$

with a special logarithmic potential, which will be introduced and discussed in the next theorem.

THEOREM 2. ([16], Theorem 2]): *For each $a > 1$ there exists a positive measure ν_a and a constant $b = b(a)$ with $1 < b < a$ such that*

$$\operatorname{supp}(\nu_a) = [b, a], \tag{2.21}$$

and the potential

$$p_a(z) := \int \log \left| \frac{z+x}{z-x} \right| d\nu_a(x) \tag{2.22}$$

satisfies

$$p_a(z) \begin{cases} = \log |z| & \text{for } z \in [b, a] \\ \\ > \log |z| & \text{for } z \in [0, b). \end{cases} \tag{2.23}$$

The potential p_a and the measure ν_a are uniquely determined by the conditions in (2.23). Further we have

$$p_a(iy) = 0 \qquad \text{for all } y \in \mathbb{R}, \tag{2.24}$$

$$a' > a \Rightarrow \|\nu_{a'}\| > \|\nu_a\| \quad \text{and} \quad b' = b(a') > b = b(a), \tag{2.25}$$

the constant $b = b(a)$ and the total variation $\|\nu_a\|$ depend continuously on a, and

$$\begin{aligned} b &\to 1, \quad \|\nu_a\| \to 0 \quad \text{as } a \to 1, \\ b &\to 2, \quad \|\nu_a\| \to \infty, \quad \nu_a([b, 2]) \to 0 \quad \text{as } a \to \infty. \end{aligned} \tag{2.26}$$

More precisely

$$\lim_{a \to \infty} \frac{1}{a} \exp \left(\pi \sqrt{\|\nu_a\|} \right) = 2. \tag{2.27}$$

and

$$b = b(a) = 2 + O(\frac{1}{a} \log a) \quad \text{as } a \to \infty. \tag{2.28}$$

In H_+ the potential p_a can be represented as

$$p_a(z) = \int \log \frac{1}{|z-x|} d(\nu_a - \omega_a)(x),$$

where ω_a is a positive measure on the imaginary axis ∂H_+ satisfying

$$\|\omega_a\| = \|\nu_a\|.$$

The two measures ν_a and ω_a are absolutely continuous and their density functions are given by

$$\frac{d\nu_a(x)}{dx} = \frac{1}{\pi C_1 x} \int_b^x \frac{(C_0^2 + t^2)dt}{\sqrt{(a^2 - t^2)^3(t^2 - b^2)}}, \quad x \in [b, a], \tag{2.29}$$

$$\frac{d\omega_a(iy)}{dy} = \frac{1}{2\pi C_1 y} \int_0^y \frac{(C_0^2 - t^2)dt}{\sqrt{(a^2 + t^2)^3(t^2 + b^2)}}, \quad y \in \mathbb{R}. \tag{2.30}$$

The constants C_0^2 and C_1 in (2.29) and (2.30) are determined by

$$C_0^2 = \frac{\displaystyle\int_0^\infty \frac{t^2\,dt}{\sqrt{(t^2+a^2)^3(b^2+t^2)}}}{\displaystyle\int_0^\infty \frac{dt}{\sqrt{(t^2+a^2)^3(b^2+t^2)}}} = \frac{a^2 E(q) - b^2 K(q)}{K(q) - E(q)}, \qquad (2.31)$$

$$C_1 = \frac{1}{\pi}\int_0^b \frac{(C_0^2+t^2)\,dt}{\sqrt{(t^2-a^2)^3(b^2-t^2)}} = \frac{K(q)E(q') + E(q)K(q') - K(q)K(q')}{\pi a[K(q) - E(q)]}$$

$$= 2a[K(q) - E(q)], \qquad (2.32)$$

where $K(q)$ and $E(q)$ are the complete elliptical integrals of the first and second kind, respectively, with moduli

$$q = \sqrt{1 - (b/a)^2}, \quad q' = b/a. \qquad (2.33)$$

Theorem 2 has been proved in [16], §3. However, the formulation given here differs by a transformation $z \mapsto 1/z$ of the independent variable and the integration variable. Corresponding tranformations have to be applied to the formulae for the constants a, b, C_0^2, and C_1. The second equality in (2.31) and (2.32) are consequences of the formulae (3.48) and (3.52) in [16], but they can also be derived directly from, 3.158.3, 3.159.3, 3.158.11, and 3.159.11. The last equality in (2.32) follows from the identity $K(q)E(q') + K(q')E(q) - K(q)K(q') = \pi/2$. For a definition of elliptical integrals see, 8.1.

The special form of the density functions (2.29) and (2.30) gives an hint to the proof of Theorem 2. It is a systematic study of the function

$$f(z) = \int_0^z \frac{1}{\zeta}\int_0^\zeta \frac{(C_0^2 + t^2)\,dt}{\sqrt{(a^2-t^2)^3(t^2-b^2)}}\,d\zeta. \qquad (2.34)$$

For an appropriated choice of the number $b = b(a)$ one gets

$$p_a(z) = \text{const. Re } f(z). \qquad (2.35)$$

We will not discuss here more details of the proof. Instead we come back to the second part of the proof of Theorem 1, which is basically a comparison of $\log|(1/4z)\Phi_n|$ with the potential p_a. From (2.17) we know that the function (2.20) can be represented as

$$q_n(z) := \log\left|\frac{1}{4z}\Phi_n(z)\right| = \tilde\varphi_n(z) + \int \log\left|\frac{z-x}{z+x}\right|d\mu_n(x), \qquad (2.36)$$

where μ_n is the positive measure introduced in (2.17). The measure μ_n is defined on $[1/2, 1/\epsilon_n]$ and has total mass $n + 1$. Function $\tilde{\varphi}_n$ is harmonic in H_+ with boundary values $\tilde{\varphi}_n(z) = \log|(1/4z)\Phi_n(z)|$ for $z \in \partial H_+$. It is a consequence of the estimates (2.14) that the function q_n and $\tilde{\varphi}_n$ satisfy the inequalities

$$u(iy) \leq q_n(iy) = \tilde{\varphi}_n(iy) \leq v(iy) \qquad \text{for} \quad y \in \mathbb{R} \qquad (2.37)$$

if the functions u and v are defined on ∂H_+ by

$$u(iy) := \begin{cases} \log\left(\frac{1}{2}\left(1 + \sqrt{1 - \frac{1}{4}y^{-2}}\right)\right) & \text{for} \quad y \in \mathbb{R} \setminus (-\frac{1}{2}, \frac{1}{2}) \\ \log\left(\frac{1}{2}\right) & \text{for} \quad y \in (-\frac{1}{2}, \frac{1}{2}) \end{cases} \qquad (2.38)$$
$$v(iy) := \log(1 + y^{-2}) \qquad \text{for} \quad y \in \mathbb{R}.$$

We shall compare q_n with a potential

$$p_n(z) := \int \log\left|\frac{z - x}{z + x}\right| d\nu_n(x), \qquad (2.39)$$

which is derived from the potential p_a introduced in (2.22) of Theorem 2 with the choice of parameter $a = 4/\epsilon_n$. The measure ν_n in (2.39) is defined in the following way: In a first step the measure $\tilde{\nu}_n$ is defined as $d\tilde{\nu}_n(x) := d\nu_{4/\epsilon_n}(4x)$. From (2.21), (2.25), and (2.26) we know that $\text{supp}(\tilde{\nu}_n) = [\tilde{b}_n, 1/\epsilon_n]$ and

$$\tilde{b}_n := \frac{1}{4}b(4/\epsilon_n) < \frac{1}{2}. \qquad (2.40)$$

The constant $b = b(a) \in (1, 2)$ has been introduced in Theorem 2 and the inequality in (2.40) follows from (2.25) and (2.26). From (2.26) together with (2.40) it further follows that

$$\lim_{n \to \infty} \tilde{\nu}_n([\tilde{b}_n, 1/2]) = 0. \qquad (2.41)$$

The measure ν_n is then defined by balayage of the measure $\tilde{\nu}_n$ out of the domain $H_+ \setminus [1/2, 1/\epsilon_n]$. From the properties of balayage together with (2.23) and (2.40) it follows that

$$p_n(z) = p_{4/\epsilon_n}(4z) = \begin{cases} 0 & \text{for } z \in \partial H_+ \\ \log\frac{1}{|4z|} & \text{for } z \in [1/2, 1/\epsilon_n] \end{cases}, \qquad (2.42)$$

and the function p_n is harmonic in $H_+ \setminus [1/2, 1/\epsilon_n]$.

By solving a Dirichlet problem we can assume that the two functions u and v introduced in (2.38) are defined and harmonic throughout H_+. In [16], Lemma 2.5, it has been shown that there exist positive measures ψ_{0n}, ψ_{1n} with

$$\|\psi_{0n}\| < 1, \quad \|\psi_{1n}\| < \frac{1}{2}, \quad \operatorname{supp}(\psi_{jn}) = [1/2, 1/\epsilon_n], \quad j = 1, 2, \qquad (2.43)$$

such that

$$
\begin{aligned}
h_{0n}(z) &:= u(z) + \int \log \left| \frac{z - x}{z + x} \right| d\psi_{0n}(x) = 0 \\
h_{1n}(z) &:= v(z) - \int \log \left| \frac{z - x}{z + x} \right| d\psi_{1n}(x) = 0
\end{aligned}
\qquad (2.44)
$$

for $z \in [1/2, 1/\epsilon_n]$, and we have

$$\psi_{0n} \xrightarrow{*} \psi_0 \quad \text{and} \quad \psi_{1n} \xrightarrow{*} \psi_1 \quad \text{as} \quad n \to \infty, \qquad (2.45)$$

where ψ_0 and ψ_1 are psitive measures on $[1/2, \infty]$ with $\|\psi_0\| < 1$ and $\|\psi_1\| < 1/2$.

After these preparations we are ready for a comparison of p_n and q_n. It follows from (2.36), (2.13), (2.37), (2.42), and (2.44) that the measures μ_n, ν_n, and psi_{jn}, $j = 1, 2$, satisfy the inequalities

$$
\begin{aligned}
(q_n - p_n - h_{0n})(z) &\begin{cases} \geq 0 & \text{for } z \in \partial H_+ \\ = 0 & \text{for } z \in [1/2, 1/\epsilon_n], \end{cases} \\
(q_n - p_n - h_{1n})(z) &\begin{cases} \leq 0 & \text{for } z \in \partial H_+ \\ = 0 & \text{for } z \in [1/2, 1/\epsilon_n]. \end{cases}
\end{aligned}
\qquad (2.46)
$$

Using the representations of the functions p_n, q_n, and h_{jn}, $j = 1, 2$, we deduce from (2.46) that the measures μ_n, ν_n, and ψ_{jn} satisfy the inequalities

$$-\psi_{1n} \leq \mu_n - \nu_n \leq \psi_{0n} \quad \text{for all } n \in \mathbb{N}. \qquad (2.47)$$

Since we know from (2.15) that $\|\mu_n\| = n + 1$, it follows with (2.43) that

$$n < \|\nu_n\| < n + 1.5 \quad \text{for all } n \in \mathbb{N}. \qquad (2.48)$$

Because of (2.41) and the definition of ν_n by balayage, we deduce from (2.48) that

$$n < \|\nu_{4/\epsilon_n}\| < n + 1.5 \quad \text{for } n \text{ sufficiently large.} \qquad (2.49)$$

The inequalities (2.49) together with the limit (2.27) in Theorem 2 allow to draw the final conclusion. We have

$$2 = \lim_{n \to \infty} \frac{\epsilon_n}{4} e^{\pi \sqrt{\|\nu_{4/\epsilon_n}\|}} = \frac{1}{4} \lim_{n \to \infty} \epsilon_n e^{\pi \sqrt{n}}. \qquad (2.50)$$

The first equality in (2.50) follows from (2.27) and the choice $a = 4/\epsilon_n$, the second one from (2.49). Thus, we have proved the limit (2.1), and the proof of Theorem 1 is completed.

In the present discussion we did not have enough space to give all details of the proofs of Theorem 1 and 2 (they can be found in [16]). We close the section by drawing attention to the inequalities in (2.47). It follows from representation (2.17) that the measure μ_n gives us very precise information about the distribution of extreme points of the error function $|x| - r_n^*(x)$ on $(0, 1]$, and for the measure ν_n we have a rather explicit representation since its density is nearly identical with formula (2.29) in Theorem 2. Hence, the inequalities (2.46) will allow us to find the assumptotic distribution of the extreme points for $n \to \infty$. More about this in the next section.

3. Asymptotic Distribution of Poles, Zeros, and Extreme Points

In order to understand better the convergence behavior of the approximants r_n^* we investigate the asymptotic distribution of its poles and zeros and also the distribution of the extreme points of the error function $|x| - r_n^*(x)$ on $[-1, 1]$. Typically, we shall give for all three distributions two versions of precision. The more precise version allows us to determine asymptotically the location of each object (pole, zero, or extreme point), while the less precise version gives only an estimate for the density of these objects. We shall start with an investigation of the extreme points.

In (2.3) and (2.4) we have seen that there are exactly $n + 2$ extreme points $x_j = x_{jn}$, $j = 0, \ldots, n+1$, on the interval $[0, 1]$ if n is even, and the first one of these points, is equal to 0. Since r_n^* is an even function, there are $n+1$ additional extreme points on $[-1, 0)$. Hence, altogether there are $2n + 3$ extreme points for even n. Because of the oscillation of the function R_n with constant amplitude on the intervall $[1/2, 1/\epsilon_n]$ (see (2.10) and (2.11)), it follows from the definition of Φ_n in (2.12) and the representation (2.17) that the measure μ_n gives a full description of the position of all extreme points on $(0, 1]$. Indeed, it follows from (2.12), (2.17), and some thoughts that the distribution function

$$F_n(x) := \mu_n([0, x/\epsilon_n]), \quad x \in [0, 1], \tag{3.1}$$

allows one to determine the extreme points x_{jn} uniquely. We have

$$F_n(x_{jn}) = j \quad \text{for} \quad j = 1, \ldots, n+1. \tag{3.2}$$

Unfortunately, we have no explicit representation for the measure μ_n, but the lower and upper bounds in (2.47) tell us that the measures μ_n and ν_n cannot differ too much, and for ν_n we can derive an approximative density function. Basically, we have to use formula (2.29) and will derive estimates for the constants C_0^2, C_1, a, and b.

THEOREM 3. ([17], Theorem 3): *For the distribution function (3.1) we have the approximative formula*

$$F_n(x) = n + 1 - \sqrt{n} \int\limits_x^1 g_n(t)dt + o(1) \quad as \quad n \to \infty, \qquad (3.3)$$

the error estimate holds uniformly for $x \in [x_0, 1]$, $0 < x_0$ *fixed, and the density function* g_n *is given by*

$$g_n(x) := \frac{2}{x\sqrt{1 - x^2}} + \frac{2}{\pi\sqrt{n}x} \log \frac{x}{1 + \sqrt{1 - x^2}}. \qquad (3.4)$$

(*By* $o(.)$ *we denote the small Landau symbol*).

Remark. Since the error term in (3.3) tends to zero for all $x \in (0, 1]$, it follows from (3.2) that the asymptotic formula is so precise that the exact number of extreme points on any given interval $[c, d] \subseteq [0, 1]$ can be determined for n sufficiently large. Hence, it is possible to determine also the individual position of extreme points.

If we drop the second term in (3.4), we get a simpler, but less precise approximation of the distribution function F_n. We then have

$$g_n(x) = \frac{2}{x\sqrt{1 - x^2}} + O(\frac{1}{\sqrt{n}}) \quad as \quad n \to \infty, \qquad (3.5)$$

and the error estimates holds unifomly for $x \in [x_0, 1]$, $0 < x_0$ fixed. This simplification gives the following

COROLLARY 4. *For every intervall* $[c, d] \subseteq (0, 1]$ *we have*

$$\lim_{n \to \infty} \frac{1}{\sqrt{n}} \text{card}\{j \mid x_{jn} \in [c, d]\} = 2 \int\limits_c^d \frac{dx}{x\sqrt{1 - x^2}}. \qquad (3.6)$$

Remark. It follows from (3.6) that the extreme points are asymptotically dense in $[-1, 1]$. On every closed subinterval of $[-1, 1] \setminus \{0\}$ the number of extreme points tends to infinity like $\sqrt{n}$ as $n \to \infty$. However, the relative frequency $\frac{1}{n}\text{card}\{j \mid x_{jn} \in [c, d]\}$ has zero density evereywhere on $[-1, 1] \setminus \{0\}$, wich implies that almost all extreme points tend to $x = 0$ as $n \to \infty$.

In Figure 3.1 below we have plotted the graph of the asymptotic distribution based on (3.6) together with the exact cummulative distribution for $n = 40$.

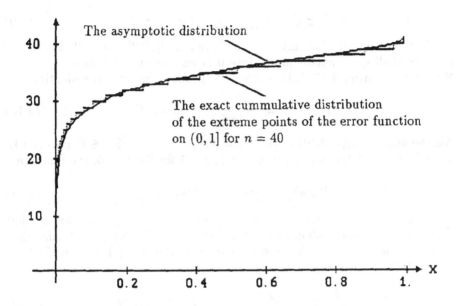

Figure 3.1

The asymptotic and the exact cummulative distribution of extreme
points of $|x| - r_n^*(x)$ on $(0, 1]$ for $n = 40$

A complete proof of Theorem 3 is contained in [17]. Here, we continue
with the discussion of results about the distribution of poles and zeros of the
approximant r_n^*. For even n there exist exactly n poles and zeros, which we
will denote by $\pi_{1n}, \dots, \pi_{nn}$ and $\zeta_{1n}, \dots, \zeta_{nn}$.

The rational function r_n defined in (2.2) allows one to obtain a simple characterisation of the location of these poles $\pi_{1n}, \dots, \pi_{nn}$ and zeros
$\zeta_{1n}, \dots, \zeta_{nn}$. It follows from (2.2) that if $z \neq 0, \infty$, then $z = \pi_{jn}$ is a pole
and $z = \zeta_{jn}$ is a zero of r_n^* if and only if

$$r_n(\pi_{jn}) = 1 \quad \text{and} \quad r_n(\zeta_{jn}) = -1, \tag{3.7}$$

respectively. Since all approximants r_n^* are even functions we can assume
without loss of generality that n is even.

LEMMA 5. *All poles and zeros of r_n^* lie on the immaginary axis $i\mathbb{R}$. Poles
and zeros are interlacing and they lie symmetric with respect to $x = 0$. They
can be numbered in such a way that*

$$0 < \frac{1}{i}\zeta_{1n} < \frac{1}{i}\pi_{1n} < \frac{1}{i}\zeta_{2n} < \frac{1}{i}\pi_{2n} < \cdots < \frac{1}{i}\zeta_{n/2,n} < \frac{1}{i}\pi_{n/2,n} \tag{3.8}$$

and

$$\zeta_{jn} = -\zeta_{(j-n/2),n}, \pi_{jn} = -\pi_{(j-n/2),n} \quad \text{for } j = n/2 + 1, \ldots, n. \tag{3.9}$$

Different proofs of lemma 5 can be found in [3], [22], and [23]. Below, we however shall give another proof since it is very short and simple.

Proof of Lemma 5: It follows from (2.2) that r_n satisfies the identity

$$|r_n(-\bar{z})| = \frac{1}{|r_n(z)|} \quad \text{for all } z \in \mathbb{C}. \tag{3.10}$$

Among other things (3.10) implies that $|r_n(iy)| = 1$ for all $y \in \mathbb{R}$ (see (2.6)). Since $|r_n(z)| < 1$ for $\text{Re}(z) > 0$ and $|r_n(z)| > 1$ for $\text{Re}(z) < 0$, the function

$$H_n(y) := -\frac{1}{\pi} \arg(r_n(iy)), \, y \in \mathbb{R}, \tag{3.11}$$

is monotonically increasing. Let the argument function be normalized so that $H_n(0) = 0$. It then follows that $H_n(-y) = -H_n(y)$ for $y \in \mathbb{R}$ and that $H_n(i\infty) = n + 1$ since r_n has $n + 1$ zeros on $(0, 1]$ and $n + 1$ poles on $[-1, 0)$. From (3.7) we know that

$$\begin{aligned} H_n(\zeta_{jn}) &= 2j - 1 \\ H_n(\pi_{jn}) &= 2j \end{aligned} \quad \text{for } j = 1, \ldots, n/2. \tag{3.12}$$

The identities (3.12) give a characterization for $n/2$ zeros and poles of r_n^* on the positive imaginary axis. Since r_n^* is a real function, the same number of zeros and poles lie symmetrically on the negative imaginary axis in accordance with (3.9). All together these are n poles and zeros, and therefore all zeros and poles of r_n^*. □

We have seen in (3.12) that the funtion H_n characterizes zeros and poles of r_n^* in the same way as the distribution function F_n in Theorem 3 has characterized the extreme points of the error function $|x| - r_n^*$. The next theorem is an analogue to Theorem 3, and it together with (3.12) allows us to determine the location of poles and zeros on the positive imaginary axis.

THEOREM 6. ([17], Theorem 4): *For the function H_n we have the approximative formula*

$$H_n(y) = n + 1 - \sqrt{n} \int_y^\infty h_n(t)dt + o(1) \quad \text{as } n \to \infty \tag{3.13}$$

the error estimate holds unifomly for $y \in [y_0, \infty]$, $0 < y_0$ fixed, and the density function h_n is given by

$$h_n(y) := \frac{1}{y\sqrt{1 + y^2}} + \frac{1}{\pi\sqrt{ny}} \log \frac{y}{1 + \sqrt{1 + y^2}}. \tag{3.14}$$

Remark.   Like in Theorem 3 so also here the precision of the asymptotic formula (3.13) is good enough to determine the exact number of poles and zeros on any given interval $[c, d] \subseteq (i0, i\infty]$ for n sufficiently large.

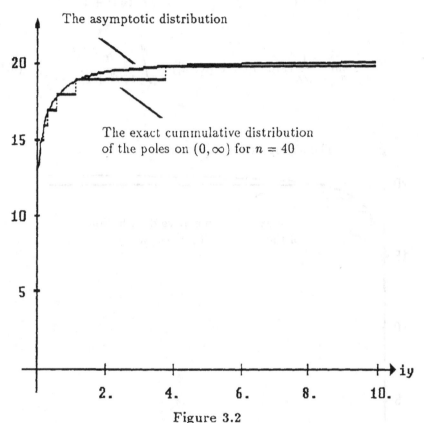

Figure 3.2

The asymptotic and the exact cummultative distribution of poles of r_n^* on $(0, \infty]$ for $n = 40$.

A simplification like in (3.5) leads to the approximative density function

$$h_n(y) = \frac{1}{y\sqrt{1 + y^2}} + O(\frac{1}{\sqrt{n}}) \quad \text{as} \quad n \to \infty \tag{3.15}$$

with an error estimate that holds uniformly for $y \in [y_0, \infty)$, $0 < y_0$ fixed, and we have

COROLLARY 7. *For any interval* $[c, d] \subseteq (0, \infty]$ *we have*

$$\lim_{n \to \infty} \frac{1}{\sqrt{n}} \operatorname{card} \{j | \pi_{jn} \in [ic, id]\} = \int_c^d \frac{dt}{t\sqrt{t^2 + 1}} \qquad (3.16)$$

and

$$\lim_{n \to \infty} \frac{1}{\sqrt{n}} \operatorname{card} \{j | \zeta_{jn} \in [ic, id]\} = \int_c^d \frac{dt}{t\sqrt{t^2 + 1}}. \qquad (3.17)$$

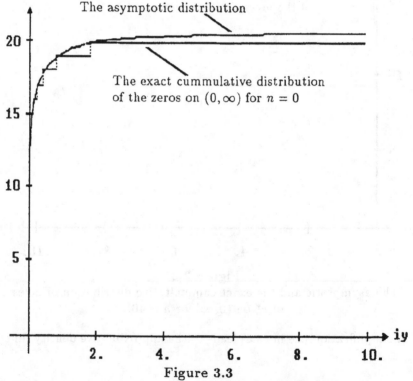

Figure 3.3

The asymptotic and the exact cummultative distribution of zeros
of r_n^* on $(0, \infty]$ for $n = 40$.

Remark. From (3.16) and (3.17) we learn that the number of poles and zeros on an imaginary intervall $[ic, id] \subseteq i\mathbb{R} \setminus \{0\}$, grows like $\sqrt{n}$ as $n \to \infty$. Especially, the result shows that poles and zeros are asymptotically dense on the imaginary axis, which confirms a numerical oberservation in [23], where it has been shown that the largest pole and zero of r_n^* tend to infinity as $n \to \infty$. Further, the result shows that almost all poles and zeros tend to $x = 0$ as $n \to \infty$.

In Figure 3.2 and 3.3 we have plotted the graphs of the asymptotic distributions of poles and zeros of r_n^* together with the exact cummulative distributions for $n = 40$.

The full proof of Theorem 6 is too long to be included in this paper. It can be found in [17]. Results obtained in the proof of Theorem 3 play an important role for Theorem 6 since there exists a relationship between the distribution of extreme points of $|x| - \nu_n^*(x)$ on $(0, 1]$ and the distribution of zeros and poles of r_n^* on $i\mathbb{R}$. The reason for that is the following. Between two adjacent extreme points on $(0, 1]$ there is exactly one zero of r_n defined in (2.2). These are altogether $n + 1$ zeros, and therefore all zeros of r_n. The zeros determine the behavior of r_n on the $i\mathbb{R}$, and because of (3.7) also the location of poles and zeros of r_n^*. See [17], §3 for more details.

4. Two Conjectures

It has already been mentioned in Introduction that the problem of rational approximation of $|x|$ on $[-1, 1]$ is closely related with that of x^α, $\alpha > 0$, on $[0, 1]$. A simple substitution $z \mapsto z^2$ shows that

$$E_{2n,2n}\left(|x|^{2\alpha}, [-1, 1]\right) = E_{nn}\left(x^\alpha, [0, 1]\right), \quad n \in \mathbb{N}. \tag{4.1}$$

Hence, rational approximation of $|x|$ on $[-1, 1]$ corresponds to rational appoximation of $\sqrt{x}$ on $[0, 1]$.

A strong error estimate for rational approximation of x^α on $[0, 1]$ completely corresponding to formula (2.1) in Theorem 1, would be

$$E_{nn}(x^\alpha, [0, 1]) = 8|\sin \pi \alpha| e^{-2\pi\sqrt{\alpha n}}(1 + o(1)) \quad \text{as } n \to \infty, \alpha > 0. \tag{4.2}$$

The author is now working on a proof for this estimate, and partial results look very promissing. We note that trivially $E_{nn}(x^\alpha, [0, 1]) = 0$ for all $\alpha \in \mathbb{N}$, and $E_{nn}(x^\alpha, [0, 1]) = \infty$ for all $n \in \mathbb{N}$ and $\alpha < 0$.

The error formula (4.2), the distribution of poles, zeros, and extreme points described in Theorem 3 and 6, and the results in [10], [7], and [22], which have been mentioned in (1.5) and (1.6), show a decisive role of the type of singularity of f at $x = 0$ for the rate of rational approximation. The exponent in the approximation formula corresponds directly to the exponent of the singularity at $x = 0$. The analytical background of the proof of Theorem 1 also suggests that this connection should hold true for more general types of functions. Based on these observations we make the following

CONJECTURE 1. *Let* $U \subseteq \mathbb{C}$ *be a neighborhood of* $[0, 1]$, *let* f *be a function that is locally analytic in* $U \setminus \{0\}$, *and assume that there exists* $\alpha > 0$ *so that*

$$\limsup_{z \to 0}(|z|^{-\alpha+\epsilon}|f(z)|) = 0 \qquad \text{and} \qquad \liminf_{z \to 0}(|z|^{-\alpha-\epsilon}|f(z)|) = \infty \quad (4.3)$$

for any $\epsilon > 0$. *Then*

$$e^{-(2\pi+\epsilon)\sqrt{\alpha n}} \leq E_{nn}(f, [0, 1]) \leq e^{-(2\pi-\epsilon)\sqrt{\alpha n}} \qquad (4.4)$$

for any $\epsilon > 0$ *and* n *sufficiently large. The lower bound holds for* $\alpha \in \mathbb{N}$ *only if* f *is not analytic at* $z = 0$.

Conjecture 1 allows one to deduce informations about the exponent of the singularity of f at $z = 0$ from the rate of rational approximability. On the other hand it shows that only the singularity at $z = 0$ is decisive for the rate. All formula remain unchanged if the intervall $[0, 1]$ is substituted by an intervall $[0, \delta]$ with $0 < \delta \leq 1$. The conjecture is backed by results in [1], where Markov functions of type z^α have been considered (see (1.6) and (1.7)).

It is not necessary to restrict the rational approximation of f to an interval. If we approximate f on a more general set E with a corner at $z = 0$, then the angle of this corner has to be taken into consideration. We make the following

CONJECTURE 2. *Let* $E \subseteq \mathbb{C}$ *be a compact, simply connected set with smooth boundary,* $0 \in \partial E$, *and let* E *have a corner with inner angle* θ, $0 \leq \theta < 2\pi$, *at* $z = 0$. *Let further* U *be a neighborhood of* E, *and let* f *be a function locally analytic in* $U \setminus \{0\}$ *satisfying (4.3) with* $\alpha > 0$. *Then*

$$e^{-(2\pi+\epsilon)\sqrt{(1-\theta/2\pi)\alpha n}} \leq E_{nn}(f, E) \leq e^{-(2\pi-\epsilon)\sqrt{(1-\theta/2\pi)\alpha n}} \qquad (4.5)$$

for any $\epsilon > 0$ *and* n *sufficiently large. The lower bound in (4.5) holds for* $\alpha \in \mathbb{N}$ *only if* f *is not analytic at* $z = 0$.

Remark. An interesting special case in Conjecture 2 is $E_{nn} := \{|z-1| \leq 1\}$, which has been considered in [1]. There we have $\Theta = \pi$.

It is well known that the problem of rational approximation is closely related with the third and fourth Zolotarev problems (see [9] for details). In connection with Conjecture 2 it may be interesting to consider a paper of S. Järner [11], where weighted versions of Zolotarev's problems have been investigated for a pair of sets with one common point containing prescribed angles at this point.

References

[1] J.-E.Anderson, *Rational approximation to functions like x^α in integral norms*, Anal. Math. **14** (1988), pp. 11–25.

[2] S. Bernstein, *Sur meilleure approximation de $|x|$ par des polynômes de degrés donnés*, Acta Math. **37** (1913), pp. 1–57.

[3] H.-P. Blatt, A. Iserles, and E.B. Saff, *Remarks on the behavior of zeros and poles of best approximating polynomials and rational functions*. In: Algorithms for Approximation (J.C. Mason and M.G. Cox, eds.), pp. 437–445. Inst. of Math. and Its Applic. Conference Series, Vol. 10, Claredon Press, Oxford, 1987.

[4] A.P. Bulanow, *The approximation of $x^{1/3}$ by rational functions*, Vesci Akad. Navuk BSSR Ser. Fiz.-Navuk 1968, **2**, pp. 47–56.(Russian)

[5] A.P. Bulanow, *Asymptotics for the least derivation of $|x|$ from rational functions*, Mat. Sb., **76** (118) (1968), pp. 288–303; English transl. in Math. USSR Sb. 5 (1968), pp. 275–290.

[6] G. Freud and J. Szabados, *Rational Approximation to x^α*, Acta Math. Acad. Sci. Hungar. **18** (1967), pp. 393–393.

[7] T. Ganelius, *Rational approximation to x^α on $[0,1]$*, Anal. Math. **5** (1979), pp. 19–33.

[8] A.A. Gonchar, *On the speed of rational approximation of continuous functions with characteristic singularities*, Mat. Sb., **73** (115) (1967), pp. 630–638; English transl. in Math USSR Sb., **2** (1967).

[9] A.A. Gonchar, *Rational approximation of the function x^α*, Constructive Theory of Functions (Proc. Internat. Conf., Varna 1970), Izdat. Bolgar. Akad. Nauk, Sofia, 1972, pp. 51–53. (Russian)

[10] A.A. Gonchar, *The rate of rational approximation and the property of single-valuedness of an analytic function in a neighborhood of an isolated singular point*, Mat. Sb., **94** (136) (1974), pp. 265–282; English transl. in Math USSR Sb., **23** (1974).

[11] S.Järner, *On weighted Zootarev problem for rational functions*, Thesis, University of Götebrg, Göteborg, 1982

[12] G. Meinardus, *Approximation of Functions: Theory and Numerical Methods*, Springer-Verlag, New York 1967.

[13] D.j. Newman, *Rational approximation to $|x|$*, Mich. Math. J., **11** (1964), pp. 11–14.

[14] T.J. Rivlin, *An Introduction to the Approximation of Functions*, Blaisdell Publ. Co., Waltham, Mass. 1969.

[15] E.B. Saff, H. Stahl, and M. Wyneken, *The Walsh table for x^α*, manuscript.

[16] H.Stahl, *Best uniform rational approximation of |x| on [-1,1]*, to appear in Mat. Sb.

[17] H.Stahl, *Poles and zeros of best rational approximants of |x|*, submitted to J.Comp.-Appl.Math.

[18] J. Tzimbalario, *Rational approximation to x^α*, J. App. Theory, 16 (1976), pp. 187-193.

[19] R.S.Varga and A.J.Carpenter, *On the Bernstein conjecture in approximation theory*, Constr.Approx., 1 (1985), pp. 333-348; Russian transl. in Mat.Sb., 129 (171) (1986), pp. 535-548.

[20] N.S. Vjacheslavov, *The approximation of |x| by rational functions*, Matem. Zametki 16 (1974), pp. 163-171, (Russian).

[21] N.S. Vjacheslavov, *On the uniform approximation of |x| by rational functions*, Dokl. Akad. Nauk SSSR, 220 (1975), pp. 512-515; English transl. in Soviet Math. Dokl., 16 (1975), pp. 100-104.

[22] N.S. Vjacheslavov, *On the approximation of x^α by rational functions*, Izv. Akad. Nauk USSR, 44 (1980), English transl. in Math. USSR Izv. 16 (1981), pp. 83-101.

[23] R.S. Varga, A. Ruttan and A.J. Carpenter, *Numerical results on best uniform rational approximation of |x| on [-1,1]*, to appear in Mat.Sb.

TFH-Berlin, Luxemburg str.

10/1000 Berlin 65

Germany

CLASSICAL
BIORTHOGONAL RATIONAL FUNCTIONS†

Mizan Rahman* and S.K. Suslov**

A general set of biorthogonal rational functions, considered previously
by Rahman and Wilson, is shown to satisfy a second-order linear difference equation of a nonuniform lattice. In the spirit of Hahn's approach
for orthogonal polynomials, raising and lowering operators as well as a
Rodriguez-type formula are obtained for these functions which contain
the classical orthogonal polynomials as limiting cases. Their biorthogonality in the discrete case is established by means of a Sturm-Liouville
type argument. An outline of Wilson's technique for representing them
as Gram determinants is also given.

1. Introduction

The important contributions of Chebyshev [7] and Hahn [10] to the theory of classical orthogonal polynomials are well-known. In recent years the
theory has been greatly enriched by the works of several authors, most importantly of R. Askey and his collaborators (see, for example, [1], [3–5], [8], [9],
[11], [12], [19–21] and the references cited therein). It is clear by now that
the characterizing properties of these polynomials are the existence of difference equations of hypergeometric type, the raising and lowering difference
operators, Rodriguez-type formulas, series expansions, duality, orthogonality
with respect to positive measures, and moment-relations. They are equivalent to each other and any one of these properties may be taken, in the
spirit of Hahn's [10] approach, as a basis for defining this class of orthogonal
polynomials.

The original family of classical orthogonal polynomials that included the
Jacobi, Laguerre and Hermite polynomials has only recently been expanded
by Andrews and Askey [1] to incorporate the polynomials found by Hahn as
well as those by Racah, Askey and Wilson, see [3–5], [21] and [23]. A subset

---

†This work, supported in part by NSERC grant #A6197, was completed while the
second author (SKS) was visiting Carleton University, Jan.-April, 1991

of this enlarged family that contains the Hahn [7], q-Hahn [10], the Racah
and q-Racah ([3], [4], [21]) polynomials has been called Charm polynomials
by Atakishiyev, Kuznetsov and Suslov [6] to epitomize the contributions of
Chebyshev, Hahn, Askey, Racah and Wilson (the W of Wilson is replaced
by its inversion M to round out a meaningful word: CHARM). Roughly
speaking, they are the polynomial solutions of a difference equation of the
form

$$\sigma(s)\frac{\nabla}{\nabla x_1(s)}\left[\frac{\Delta y(s)}{\Delta x(s)}\right] + \tau(s)\frac{\Delta y(s)}{\Delta x(s)} + \lambda_n y(s) = 0, \qquad (1.1)$$

where $\sigma(s)$ and $\tau(s)$ are certain functions of s, and the lattice $x(s)$ is given
by

$$x(s) = \begin{cases} C_1 q^{-s} + C_2 q^s, & q \neq 1, \\ C_1 s^2 + C_2 s, & q = 1, \end{cases} \qquad (1.2)$$

λ_n is the eigenvalue that depends on the degree n of the polynomial solution
but not on s, and the forward and backward differences are defined, as usual,
by $\Delta f(s) = f(s+1) - f(s)$, $\nabla f(s) = \Delta f(s-1)$. Also $x_\mu(s) = x(s + \mu/2)$, $s, \mu \in \mathbb{C}$. It is possible to add an arbitrary constant on the right side of
(1.2) but since it doesn't play any essential role in the final analysis we set it
to be 0. For the Charm polynomials $p_n(x) = p_n^{(\alpha,\beta)}(x; a, b)$, the coefficients
$\sigma(s)$ and $\tau(s)$ in (1.1) are taken in such a way that the following factorizations
hold:

$$\begin{aligned} \sigma(s) &= [x_{\alpha-1}(b) - x_{\alpha-1}(s - \alpha)][x_{-\beta-1}(s) - x_{-\beta-1}(a)], \\ \sigma(s) &+ \tau(s) \nabla x_1(s) \\ &= [x_{\alpha-1}(b) - x_{\alpha-1}(s+1)][x_{1-\beta}(s+\beta) - x_{1-\beta}(a-1)]. \end{aligned} \qquad (1.3)$$

The weight function for the Charm polynomials satisfies the Pearson–type
equation

$$\Delta[\rho(s)\sigma(s)] = \rho(s)\tau(s) \nabla x_1(s), \qquad (1.4)$$

and is taken as

$$\rho(s) = x_{\alpha-1}(b) - x_{\alpha-1}(s)]^{(\alpha)}[x(s) - x(a - 1)]^{(\beta)}, \qquad (1.5)$$

where $[x_\nu(s) - x_\nu(z)]^\mu$ is the generalized power defined by

$$\begin{aligned} [x_\nu(s) - x_\nu(z)]^{(\mu)} &= [x_\nu(s) - x_\nu(z)][x_\nu(s) - x_\nu(z - 1)]^{(\mu-1)} \\ &= [x_\nu(s) - x_\nu(z)]^{(\mu-1)}[x_\nu(s) - x_\nu(z - \mu + 1)], \end{aligned} \qquad (1.6)$$

$$[x_\nu(s) - x_\nu(z)]^{(\mu)} = [x_{\nu-1}(s+1) - x_{\nu-1}(z)]^{(\mu-1)}[x_{\nu-\mu+1}(s)$$

$$- x_{\nu-\mu+1}(z)] = [x_{\nu-\mu+1}(s+\mu-1) - x_{\nu-\mu+1}(z)] \quad (1.7)$$

$$\cdot [x_{\nu-1}(s) - x_{\nu-1}(z)]^{(\mu-1)},$$

see, for example, [20]. We should like to point out that the properties (1.6) and (1.7) are not satisfied by any general lattice $x(s)$ but only by the one defined by (1.2). Also, note an analogy between (1.5) and the weight function $\rho(x) = (1-x)^\alpha(1+x)^\beta$ for the Jacobi polynomials.

The discrete orthogonality relation for the Charm polynomials has the form

$$\sum_{s=a}^{b-1} p_m^{(\alpha,\beta)}(x(s))p_n^{(\alpha,\beta)}(x(s))\rho(s) \bigtriangledown x_1(s) = d_n^2 \delta_{m,n}, \quad (1.8)$$

where the boundary conditions are chosen to be

$$\sigma(a) = 0,$$
$$\sigma(b-1) + \tau(b-1) \bigtriangledown x_1(b-1) = 0. \quad (1.9)$$

The purpose of this paper is to extent this class of orthogonal polynomials in much the same way as Rahman [13], [14] and Wilson [22], [23] extended the classical orthogonal polynomials. In fact, the biorthogonal rational functions that we shall consider here are essentially the same as those found by Rahman and Wilson. However our approach will be different from theirs. In the following sections we show that a slight change in the expressions on the right hand sides of (1.3) leads to a completely new phenomenon. Polynomials turn into rational functions, orthogonality into biorthogonality, hypergeometric difference equation (1.1) into one with n-dependent coefficients. We employ the same general technique as Hahn [10] did to expand the classical orthogonal polynomials. So it looks appropriate to call these new objects as classical biorthogonal rational functions. The word classical is used here not in the usual sense of time-frame, but in the sense of its content and spirit.

2. Pearson–type equation

In view of the lattice (1.2) and the factorizations (1.3) it is convenient to introduce two auxilary functions $\tilde{\sigma}(x(s))$ and $\tilde{\tau}(x(s))$ such that

$$\tau(s) = \tilde{\tau}(x(s)),$$
$$\sigma(s) = \tilde{\sigma}(x(s)) - \frac{1}{2}\tilde{\tau}(x(s)) \bigtriangledown x_1(s). \quad (2.1)$$

The Charm polynomials $\tilde{\sigma}(x(s))$ and $\tilde{\tau}(x(s))$ are polynomials of degree 2 and 1, respectively, in $x(s)$. Note that if $x(-s-\nu) = x(s)$ then (2.1) gives

$$\sigma(-s-\nu) = \sigma(s) + \tau(s) \bigtriangledown x_1(s) \quad (2.2)$$

so that the second factorization in (1.3) takes place automatically provided the first holds. The point of departure for the biorthogonal rational functions is that we now take

$$\tilde{\sigma}(x(s)) = \frac{\tilde{p}_3(x(s))}{x(s_\infty) - x(s)}, \quad \tilde{r}(x(s)) = \frac{\tilde{p}_1(x(s))}{x(s_\infty) - x(s)}, \tag{2.3}$$

where $\tilde{p}_3(x(s))$ and $\tilde{p}_1(x(s))$ are polynomials of degrees 3 and 1, respectively. Thus

$$\sigma(s) = \frac{\Sigma(s)}{x(s_\infty) - x(s)}, \quad \tau(s) = \frac{T(s)}{x(s_\infty) - x(s)}, \tag{2.4}$$

where

$$\Sigma(s) = \tilde{p}_3(x(s)) - \frac{1}{2}\tilde{p}_1(x(s)) \nabla x_1(s), T(s) = \tilde{p}_1(x(s)). \tag{2.5}$$

The crucial assumption that we make here is that

$$\Sigma(s_\infty) = 0. \tag{2.6}$$

It means that $\sigma(s)$ has no poles for linear or q-linear lattices, and has one simple pole for quadratic or q-quadratic lattices. This leads to a certain balance relation between the parameters that facilitates the second factorization in (1.3) and was found to be very important in [18] for evaluating some sums and integrals over the weight function. Note that we need a simultaneous factorization of $\Sigma(s)$ in (2.5) and of

$$\Sigma(s) + T(s) \nabla x_1(s) = \tilde{p}_3(x(s)) + \frac{1}{2}\tilde{p}_1(x(s)) \nabla x_1(s). \tag{2.7}$$

Replacing the parameter s_∞ by $c - 1$ and making use of

$$\begin{aligned} \Sigma(s) = [x_{\alpha-1}(b) - x_{\alpha-1}(s - a)][x_{-\beta-1}(s) - x_{-\beta-1}(a)] \\ \cdot [x_{\alpha+\beta}(c) - x_{\alpha+\beta}(s + 1)], \end{aligned} \tag{2.8}$$

and (2.6) one can show that

$$\begin{aligned} \Sigma(s) + T(s) \nabla x_1(s) = [x_{\alpha-1}(b) - x_{\alpha-1}(s + 1)] \\ \cdot [x_{1-\beta}(s + \beta) - x_{1-\beta}(a - 1)][x_{\alpha+\beta}(c) - x_{\alpha+\beta}(s - \alpha - \beta - 1)]. \end{aligned} \tag{2.9}$$

Since $\sigma(s) = \Sigma(s)/[x(c-1) - x(s)]$ one can show by a straightforward calculation that the solution of the Pearson–type equation (1.4) is given by

$$\begin{aligned} \rho(s) &= \rho^{(\alpha,\beta)}(s; a, b, c) \\ &= \frac{[x_{\alpha-1}(b) - x_{\alpha-1}(s)]^{(\alpha)}[x(s) - x(a - 1)]^{(\beta)}}{[x_{\alpha+\beta+1}(c) - x_{\alpha+\beta+1}(s)]^{(\alpha+\beta+2)}}. \end{aligned} \tag{2.10}$$

This solution is unique up to a periodic factor of period 1.

Rewriting (1.2) in the form

$$x(s) = \begin{cases} C_1(q^{-s} + q^{\nu+s}), & q^\nu = C_2/C_1, q \neq 1, \\ C_1 s(s + \nu), & \nu = C_2/C_1, q = 1, \end{cases} \tag{2.11}$$

we find that $\tau(s)$ can be expressed as

$$\tau(s) = A + B\frac{x(a) - x(s)}{x(1 - c - \nu) - x(s)}, \tag{2.12}$$

where

$$A = A(\alpha, \beta; a, b, c) = \frac{-\Sigma(-a - \nu)}{[x(a) - x(c - 1)] \triangledown x_1(a)},$$
$$B = B(\alpha, \beta; a, b, c) = \frac{\Sigma(1 - c - \nu)}{[x(a) - x(c - 1)] \triangledown x_1(c - 1)}, \tag{2.13}$$

and that (1.4) can be written as

$$\frac{\triangledown}{\triangledown x_1(s)}[p(s + 1)\sigma(s + 1)]$$

$$= A\rho(s)\left[1 + \frac{\alpha(\frac{a-c}{2} + 1)\gamma(\frac{a-c}{2} + 1)\gamma(a - c)\gamma(b - c)}{\gamma(1)\alpha(\frac{a-c}{2})\gamma(\frac{a-c}{2})\gamma(a - b + 1)} \right. \tag{2.14}$$

$$\left. \cdot \frac{\delta(1 - b - c - \alpha - 2\nu)\gamma(a - c - \beta)\gamma(-1)\gamma(\alpha + \beta + 2)}{\gamma(\beta + 1)\delta(a + b + \alpha)\gamma(a - c + 2)\gamma(a - c - \alpha - \beta - 1)}\phi_1(s)\right]$$

with

$$\phi_1(s) = \frac{x(a) - x(s)}{x(1 - c - \nu) - x(s)}, \tag{2.15}$$

and

$$\alpha(u) = \begin{cases} \frac{1}{2}(q^{u/2} + q^{-u/2}), & q \neq 1, \\ 1, & q = 1, \end{cases}$$

$$\gamma(u) = \begin{cases} \frac{q^{u/2} - q^{-u/2}}{q^{1/2} - q^{-1/2}}, & q \neq 1, \\ u, & q = 1, \end{cases} \tag{2.16}$$

$$\delta(u) = \triangledown x_1\left(\frac{u}{2}\right).$$

Note that we have used the same symbol to denote the parameter α and the function $\alpha(u)$. It is hoped that this will not cause any confusion for the reader.

3. The natural variables and two difference operators

The form of $\tau(s)$ in (2.12) indicates that it is linear in

$$\eta(s) = \eta(s; a, c)$$

$$= [x(a) - x(1 - c - \nu)]^{-1} \frac{x(a) - x(s)}{x(1 - c - \nu) - x(s)}. \tag{3.1}$$

The calculus of finite differences that we need to employ in this work can therefore be simplified if we use this as a natural variable instead of the lattice $x(s)$. A second variable that is equally useful is given by

$$\zeta(s) = \eta(s; a, -\alpha - \beta - c - \nu - 1)$$

$$= [x(a) - x(\alpha + \beta + c + 2)]^{-1} \frac{x(a) - x(s)}{x(\alpha + \beta + c + 2) - x(s)}. \tag{3.2}$$

Clearly, these are linear fractional transformations analogous to those well–known in analysis of ordinary hypergeometric functions $F(a, b; c; x)$. Defining $\eta_k(s) = \eta(s + \frac{k}{2}; a + \frac{k}{2}, c - \frac{3k}{2})$, $k = 0, 1, 2, \ldots$, we find that

$$\frac{\nabla \eta_k(s)}{\nabla x_k(s)} = \frac{1}{[x_k(k + 1 - c - \nu) - x_k(s)]^{(2)}}. \tag{3.3}$$

$$\frac{\nabla \zeta(s)}{\nabla x(s)} = \frac{1}{[x(\alpha + \beta + c + 2) - x(s)]^{(2)}}, \tag{3.4}$$

$$\frac{\nabla \zeta_k(s)}{\nabla x_k(s)} = \frac{1}{[x_k(\alpha + \beta + c + k - 1) - x_k(s)]^{(2)}} \tag{3.5}$$

with $\zeta_k(s) = \eta(s + \frac{k}{2}; a + \frac{k}{2}, 2 - \alpha - \beta - c - \nu - \frac{3k}{2})$, $k = 1, 2 \ldots$. If we now generalize (2.15) to

$$\phi_k(s) = \phi_k(s; a, 1 - c - \nu)$$

$$= \prod_{j=0}^{k-1} \frac{x(a + j) - x(s)}{x(1 - c - \nu + j) - x(s)} = \frac{\gamma(a - s)_k \delta(a + s)_k}{\gamma(1 - c - \nu - s)_k \delta(1 - c - \nu + s)_k}, \tag{3.6}$$

where

$$\gamma(u)_k = \prod_{j=0}^{k-1} \gamma(j + u), \delta(w)_k = \prod_{j=0}^{k-1} \nabla x_1 \left(\frac{w + j}{2} \right), \tag{3.7}$$

then we can show that

$$\frac{\nabla \phi_k(s)}{\nabla \eta(s)} = \gamma(k) \gamma(a + c + \nu - 1) \delta(k + a - c - \nu) \phi_{k-1}(s + \frac{1}{2}; a + \frac{1}{2}, \frac{5}{2} - c - \nu) \tag{3.8}$$

and that

$$\left[\frac{\nabla}{\nabla \eta_{k-1}} \cdots \frac{\nabla}{\nabla \eta_1} \frac{\nabla}{\nabla \eta}\right] \phi_k(s) = \gamma(1)_k \gamma(a+c+\nu-k)_k \delta(k+a-c-\nu)_k. \quad (3.9)$$

Putting

$$\psi_k(s) = \phi_k(s - \frac{1}{2}; a + \frac{1}{2}, \frac{3}{2} - c - \nu) \qquad (3.10)$$

and using (2.8) and (2.10) we find that

$$\rho(s)\sigma(s)\psi(s)$$

$$= (-1)^k \frac{[x_{\alpha-1}(b) - x_{\alpha-1}(s)]^{(\alpha+1)}[x_{-1}(s) - x_{-1}(a+k)]^{(\beta+k+1)}}{[x_{\alpha_s+k}(c-k) - x_{\alpha+\beta+k}(s)]^{(\alpha+\beta+2+k)}} \qquad (3.11)$$

which is the same as $\rho(s)\sigma(s)$ with a, c, β replaced by $a+k, c-k$ and $\beta+k$, respectively. Since

$$\rho^{(\alpha,\beta+k)}(s; a+k, b, c-k)$$

$$= \frac{[x_{\alpha-1}(b) - x_{\alpha-1}(s)]^{(\alpha)}[x(s) - x(a+k-1)]^{(\beta+k)}}{[x_{\alpha+\beta+1+k}(c-k) - x_{\alpha+\beta+1+k}(s)]^{(\alpha+\beta+2+k)}} \qquad (3.12)$$

$$= (-1)^k \phi_k(s)\rho^{(\alpha,\beta)}(s; a, b, c),$$

we get from (1.4) and (2.12) the following formula:

$$\frac{\Delta[\rho(s)\sigma(s)\psi_k(s)]}{\nabla x_1(s)} = \rho(s)[A_k\phi_{k+1}(s)], \qquad (3.13)$$

where

$$A_k = A(\alpha, \beta+k; a+k, b, c-k), \quad B_k = B(\alpha, \beta+k; a+k, b, c-k). \quad (3.14)$$

Using of (3.5) with $k = 1$ we can rewrite (3.13) in a more clear form

$$\frac{\nabla[\tilde{p}_1(s)\psi_k(s+1)]}{\nabla \zeta_1(s)} = \tilde{p}(s)[A_k\phi_k(s) + B_k\phi_{k+1}(s)], \qquad (3.15)$$

where

$$\tilde{\rho}(s) = \tilde{\rho}^{(\alpha,\beta)}(s; a, b, c)$$

$$= [x_1(\alpha+\beta+c) - x_1(s)]^{(2)}\rho^{(\alpha,\beta)}(s; a, b, c)$$

$$= \frac{[x_{\alpha-1}(b) - x_{\alpha-1}(s)]^{(\alpha)}[x(s) - x(a-1)]^{(\beta)}}{[x_{\alpha+\beta-1}(c) - x_{\alpha+\beta-1}(s)]^{(\alpha+\beta)}}, \qquad (3.16)$$

and

$$\tilde{p}_n(s) = \tilde{\rho}^{(\alpha+n,\beta+n)}\left(s + \frac{n}{2}; a + \frac{n}{2}, b - \frac{n}{2}, c - \frac{n}{2}\right), \quad n = 1, 2, \ldots. \quad (3.17)$$

4. Rational functions and Rodriguez-type formula

The two operators introduced in (3.8) and (3.15) indicate that they will act on a series of the form $\Sigma a_k \phi_k(s)$ in much the same way as the differential operator d/dx acts on a power series $\Sigma a_k x^k$. In fact, formula (2.14) provides a hint of what kind of series one can expect.

Let

$$
u_n(s) = u_n^{(\alpha,\beta)}(s; a, b, c)
$$

$$
= C_n \sum_{k=0}^{n} \frac{\alpha(\frac{a-c}{2}+1)_k \gamma(\frac{a-c}{2}+1)_k \gamma(b-c)_k}{\gamma(1)_k \alpha(\frac{a-c}{2})_k \gamma(\frac{a-c}{2})_k \gamma(\alpha-b+1)_k}
$$

$$
\cdot \frac{\delta(1-b-c-\alpha-2\nu)_k \gamma(a-c-\beta)_k \gamma(-n)_k \gamma(\alpha+\beta+n+1)_k}{\gamma(\beta+1)_k \delta(a+b+\alpha)_k \gamma(a-c+n+1)_k \gamma(a-c-\alpha-\beta-n)_k} \phi_k(s),
$$

(4.1)

and

$$
v_n(s) = v_n^{(\alpha,\beta)}(s; a, b, c) = u_n^{(\alpha,\beta)}(s; a, b, -\alpha-\beta-c-\nu-1),
$$

(4.2)

where

$$
C_n = (-1)^n \frac{\gamma(a-b+1)_n \gamma(\beta+1)_n}{\gamma(1)_n}
$$

$$
\cdot \frac{\delta(a+b+\alpha)_n \gamma(c-a+\alpha+\beta+1)_n}{\gamma(a-c+1)_n}.
$$

(4.3)

Corresponding to the lattices $x(s) = s, s^2, q^{-s}, C_1(q^{-s}+q^{\nu+s})$ these functions lead, respectively, to Rahman's $_4F_3$ functions [13,14], Wilson's $_9F_8$ functions [22,23], and to their q-analogues, namely, the $_4\phi_3$ and $_{10}\phi_9$ rational functions (for the meaning of these symbols see the original references or [9]).

Making use of (3.10) and (3.15) allows us to derive the following formula

$$
\frac{\nabla}{\nabla \zeta_1(s)} \left[\tilde{\rho}_1(s) u_{n-1}^{(\alpha+1,\beta+1)} \left(s + \frac{1}{2}; a + \frac{1}{2}, b - \frac{1}{2}, c - \frac{1}{2} \right) \right]
$$

$$
= -\gamma(n) \tilde{\rho}(s) u_n^{(\alpha,\beta)}(s; a, b, c).
$$

(4.4)

Putting

$$
u_{n-k}^{(k)}(s) = u_{n-k}^{(\alpha+k,\beta+k)} \left(s + \frac{k}{2}; a + \frac{k}{2}, b - \frac{k}{2}, c - \frac{k}{2} \right)
$$

(4.5)

one can deduce from (4.4) that

$$
\tilde{\rho}_k(s) u_{n-k}^{(k)}(s) = \frac{-1}{\gamma(n-k)} \frac{\nabla}{\nabla \zeta_{k+1}(s)} \left[\tilde{\rho}_{k+1}(s) u_{n-k-1}^{(k+1)}(s) \right],
$$

(4.6)

$k = 0, 1, \ldots, n - 1$. From (4.4) and (4.6) we deduce a Rodriguez–type formula

$$u_n^{(\alpha,\beta)}(s; a, b, c) = \frac{(-1)^n}{\gamma(1)_n \tilde{\rho}(s)} \nabla_s^{(n)} [\tilde{\rho}_n(s)], \tag{4.7}$$

where

$$\nabla_s^{(n)} = \frac{\nabla}{\nabla\zeta_1(s)} \frac{\nabla}{\nabla\zeta_2(s)} \cdots \frac{\nabla}{\nabla\zeta_n(s)}. \tag{4.8}$$

Also, from (3.8) and (4.1) we obtain

$$\frac{\Delta}{\Delta\eta(s)} u_n^{(\alpha,\beta)}(s; a, b, c)$$
$$= \gamma(\alpha + \beta + n + 1)[x_{\beta_1}(c) - x_{\beta_1}(a - \beta)][x_{\alpha-1}(c) - x_{\alpha-1}(b)] \tag{4.9}$$
$$\cdot u_{n-1}^{(\alpha+1,\beta+1)}(s + \frac{1}{2}; a + \frac{1}{2}, b + \frac{1}{2}, c - \frac{3}{2}).$$

Iterating (4.9) k times we get

$$\left[\frac{\Delta}{\Delta\eta_k} \frac{\Delta}{\Delta\eta_{k-1}} \cdots \frac{\Delta}{\Delta\eta_1} \frac{\Delta}{\Delta\eta}\right] u_n^{(\alpha,\beta)}(s; a, b, c)$$
$$= \gamma(\alpha + \beta + n + 1)_{k+1} \gamma(c - a + \beta - k)_{k+1} \gamma(c - b - k)_{k+1}$$
$$\cdot \delta(c + a - 1 - k)_{k+1} \delta(b + c + \alpha - 1 - k)_{k+1} \tag{4.10}$$
$$\cdot u_{n-k-1}^{(\alpha+k+1,\beta+k+1)} \left(s + \frac{k+1}{2}; a + \frac{k+1}{2}, b - \frac{k+1}{2}, c - \frac{3(k+1)}{2}\right),$$

$k = 0, 1, \ldots, n - 1$, which for $k = n - 1$ gives

$$\left[\frac{\Delta}{\Delta\eta_{n-1}} \frac{\Delta}{\Delta\eta_{n-2}} \cdots \frac{\Delta}{\Delta\eta_1} \frac{\Delta}{\Delta\eta}\right] u_n^{(\alpha,\beta)}(s; a, b, c)$$
$$= \gamma(\alpha + \beta + n + 1)_n \gamma(c - a + \beta + 1 - n)_n \gamma(c - b - n + 1)_n \tag{4.11}$$
$$\delta(c + a - n)_n \delta(b + c + \alpha - n)_n.$$

One can show similarly that

$$\frac{\Delta}{\Delta\zeta(s)} v_n^{(\alpha,\beta)}(s; a, b, c)$$
$$= \gamma(\alpha + \beta + n + 1)[x_{\beta-1}(\alpha + c + 2) - x_{\beta-1}(a - \beta)]$$
$$\cdot [x_{\alpha-1}(\beta + c + 2) - x_{\alpha-1}(b)]$$
$$\cdot v_{n-1}^{(\alpha+1,\beta+1)}(s + \frac{1}{2}; a + \frac{1}{2}, b - \frac{1}{2}, c - \frac{1}{2}) \tag{4.12}$$
$$= \gamma(\alpha + b + n + 1)\delta(a + c + \alpha + 1)\gamma(\alpha + \beta + c - a + 2)$$
$$\cdot \delta(\alpha + \beta + b + c + 1)\gamma(\beta + c + 2 - b)\cdot$$
$$\cdot v_{n-1}^{(\alpha+1,\beta+1)}(s + \frac{1}{2}; a + \frac{1}{2}, b - \frac{1}{2}, c - \frac{1}{2}).$$

Formulas similar to (4.10) and (4.11) for $v_n(s)$ can easily be proved.

Using (4.4) and (4.9) one can then show that $u_n(s)$ satisfies the following second-order difference equation

$$\sigma_n(s)\frac{\nabla}{\nabla x_1(s)}\left[\frac{\Delta u_n(s)}{\Delta x(s)}\right] + \tau_n(s)\frac{\Delta u_n(s)}{\Delta x(s)} + \lambda_n u_n(s) = 0, \qquad (4.13)$$

where

$$\sigma_n(s) = \gamma(s-a)\delta(s+a-\beta-1)\gamma(b-s+\alpha)\delta(b+s-1)$$

$$\qquad (4.14)$$

$$\cdot \gamma(c-s-1)_2\delta(c+s+\alpha+\beta+n)\delta(c+s-n-1),$$

$$\tau_n(s) = \frac{\sigma_n(-s-\nu) - \delta_n(s)}{\nabla x_1(s)}, \qquad (4.15)$$

and

$$\lambda_n = \gamma(n)\gamma(\alpha+\beta+n+1)\gamma(c-b)\delta(c+b+\alpha-1)$$

$$\cdot \gamma(c-a+\beta)\delta(c+a-1). \qquad (4.16)$$

Note that in contrast to the classical and the Charm orthogonal polynomials the coefficients σ_n and τ_n in (4.13) depend on n, n being the degree of the rational function $u_n(s)$. A similar difference equation also holds for $v_n(s)$ with appropriate change of the parameter c.

5. Biorthogonality

From (4.4) we have

$$\Delta[\tilde{\rho}_1(s-1)u_{n-1}^{(1)}(s-1)] + \gamma(n)\tilde{\rho}(s)u_n(s)\nabla\zeta_1(s) = 0$$

and hence

$$v_m(s)\Delta[\tilde{\rho}_1(s-1)u_{n-1}^{(1)}(s-1)] + \gamma(n)\tilde{\rho}(s)v_m(s)\nabla\zeta_1(s) = 0 \qquad (5.1)$$

Using

$$\Delta[f(s-1)g(s)] = f(s)\Delta g(s) + g(s)\nabla f(s)$$

with $f(s) = v_m(s)$ and $g(s) = \tilde{\rho}_1(s-1)u_{n-1}^{(1)}(s-1)$, we have

$$v_m(s)\Delta[\tilde{\rho}_1(s-1)u_{n-1}^{(1)}(s-1)]$$

$$= \Delta[v_m(s-1)\tilde{\rho}_1(s-1)u_{n-1}^{(1)}(s-1)]$$

$$- \tilde{\rho}_1(s-1)u_{n-1}^{(1)}(s-1)\nabla v_m(s) \qquad (5.2)$$

$$= \Delta[v_m(s-1)\tilde{\rho}_1(s-1)u_{n-1}^{(1)}(s-1)]$$

$$- \gamma(\alpha+\beta+m+1)k(c)\tilde{\rho}_1(s-1)u_{n-1}^{(1)}(s-1)v_{m-1}^{(1)}(s-1)\nabla\zeta(s),$$

by (4.12), where

$$k(c) = \delta(a+c+\alpha+1)\delta(\alpha+\beta+b+c+1)\gamma(\alpha+\beta+c-a+2)\gamma(\beta+c+2-b). \quad (5.3)$$

Then (5.1) yields

$$\Delta[\tilde{\rho}_1(s-1)v_m(s-1)u_{n-1}^{(1)}(s-1)]$$
$$= \gamma(\alpha+\beta+m+1)k(c)\tilde{\rho}_1(s-1)v_{m-1}^{(1)}(s-1)u_{n-1}^{(1)}(s-1)\nabla\zeta(s) \quad (5.4)$$
$$- \gamma(n)\tilde{\rho}(s)v_m(s)u_n(s)\nabla\zeta_1(s).$$

Let us assume that $b - a$ is a positive integer and that the functions involved take values in the discrete set $s = a, a+1, \dots, b-2, b-1$. Then summing the basic identity (5.4) from a to b we get

$$\tilde{\rho}_1(b-1)v_m(b-1)u_{n-1}^{(1)}(b-1) - \tilde{\rho}_1(a-1)v_m(a-1)u_{n-1}^{(1)}(a-1)$$
$$= \gamma(\alpha+\beta+m+1)k(c)\sum_{s=a}^{b-1}\tilde{\rho}_1(s-1)v_m^{(1)}(s-1)u_{n-1}^{(1)}(s-1)\nabla\zeta(s) \quad (5.5)$$
$$- \gamma(n)\sum_{s=a}^{b-1}\tilde{\rho}(s)v_m(s)u_n(s)\nabla\zeta_1(s).$$

We shall use the simplest boundary conditions

$$\tilde{\rho}_1(a-1) = \tilde{\rho}_1(b-1) = 0 \quad (5.6)$$

Formula (5.5) then leads to

$$\sum_{s=a}^{b-1} v_m(s)u_n(s)\tilde{\rho}(s)\nabla\zeta_1(s)$$
$$= \frac{\gamma(\alpha+\beta+m+1)k(c)}{\gamma(n)}\sum_{s=a}^{b-2}v_{m-1}^{(1)}(s)u_{n-1}^{(1}(s))\tilde{\rho}_1(s)\Delta\zeta(s). \quad (5.7)$$

By (3.5) and (3.16),

$$\tilde{\rho}(s)\nabla\zeta_1(s) = \rho(s)\nabla x_1(s), \quad (5.8)$$
$$\tilde{\rho}_1(s)\Delta\zeta(s) = \rho_1(s)\nabla x_2(s), \quad (5.9)$$

and we may rewrite (5.7) in the form

$$\sum_{s=a}^{b-1} v_m(s)u_n(s)\rho(s)\nabla x_1(s)$$
$$= k(c)\frac{\gamma(\alpha+\beta+m+1)}{\gamma(n)}\sum_{s=a}^{b-2}v_{m-1}^{(1)}(s)u_{n-1}^{(1)}(s)\rho_1(s)\nabla x_2(s). \quad (5.10)$$

The translations $n \to n-k$, $m \to m-k$, $\alpha \to \alpha+k$, $\beta \to \beta+k$, $a \to a+\frac{k}{2}$, $b \to b-\frac{k}{2}$, $c \to c-\frac{k}{2}$ then give

$$\sum_{s=a}^{b-k-1} v_{m-k}^{(k)}(s)u_{n-k}^{(k)}(s)\rho_k(s)\nabla x_{:+1}(s)$$

$$= k_k(c)\frac{\gamma(m+\alpha+\beta+k+1)}{\gamma(n-k)} \tag{5.11}$$

$$\cdot \sum_{s=a}^{b-k-2} v_{m-k-1}^{(k+1)}(s)u_{n-k-1}^{(k+1)}(s)\rho_{k+1}(s)\nabla x_{k+2}(s),$$

where

$$k_k(c) = \delta(a+c+\alpha+1+k)\delta(\alpha+\beta+b+c+k+1)$$
$$\gamma(\alpha+\beta+c-a+2+k)\gamma(\beta+c+2-b+k). \tag{5.12}$$

It is clear from (3.16) and (3.17) that

$$\rho_k(s) = \rho^{(\alpha+k,\beta+k)}\left(s+\frac{k}{2}; a+\frac{k}{2}, b-\frac{k}{2}, c-\frac{k}{2}\right), \quad k = 0,1,\dots .$$

From (5.10) and (5.11) we then find that

$$\sum_{s=a}^{b-1} v_m(s)u_n(s)\rho(s)\nabla x_1(s)$$

$$= k(c)\frac{\gamma(\alpha+\beta+m+1)}{\gamma(n)}\sum_{s=a}^{b-2} v_{m-1}^{(1)}(s)u_{n-1}^{(1)}(s)\rho_1(s)\nabla x_2(s)$$

$$= k(c)k_1(c)\frac{\gamma(\alpha+\beta+m+1)\gamma(\alpha+\beta+m+2)}{\gamma(n)\gamma(n-1)}$$

$$\cdot \sum_{s=a}^{b-3} v_{m-2}^{(2)}(s)u_{n-2}^{(2)}(s)\rho_2(s)\nabla x_3(s) \tag{5.13}$$

$$\vdots$$

$$= \begin{cases} 0 & \text{if } m < n \\ \prod_{k=0}^{n-1} k_k(c)\frac{\gamma(\alpha+\beta+n+1+k)}{\gamma(n-k)}\sum_{s=a}^{b-1-n}\rho_n(s)\nabla x_{n+1}(s), & \text{if } m = n. \end{cases}$$

By symmetry, we have the biorthogonality relation

$$\sum_{s=a}^{b-1} v_m^{(\alpha,\beta)}(s;a,b,c)u_n^{(\alpha,\beta)}(s;a,b,c)\rho(s)\nabla x_1(s) = d_n^2\delta_{m,n}, \tag{5.14}$$

where

$$d_n^2 = \prod_{k=0}^{n-1} k_k(c) \frac{\gamma(\alpha + \beta + n + 1 + k)}{\gamma(n-k)} \sum_{s=a}^{b-1-n} \rho_n(s) \nabla x_{n+1}(s). \qquad (5.15)$$

For different lattices the sum on the right hand side can be computed by different summation formulas in the theory of hypergeometric or basic hypergeometric series (see, for example, [9]). Its exact value is of no particular interest for our purposes.

6. Gram determinants

The rational functions $u_n(s)$ and $v_n(s)$ discussed in the previous sections also admit representations in terms of the Gram determinants, as it was evident from Wilson's [22] work. First, let us introduce the rational functions

$$f_k(s) = \prod_{i=0}^{k-1} \frac{x(b + \alpha + i) - x(s)}{x(c - 1 - i) - x(s)},$$
$$g_j(s) = \prod_{i=0}^{j-1} \frac{x(s) = x(a - \beta - 1 - i)}{x(\alpha + \beta + c + 2 + i) - x(s)}, \qquad (6.1)$$

$f_0(s) = g_0(s) = 1$. These functions are similar to, but not quite the same as, $\phi_k(s)$ and $\psi_k(s)$ defined in section 3. It follows from (2.10) that

$$g_j(s) f_k(s) \rho^{(\alpha,\beta)}(s; a, b, c) = \rho^{(\alpha+k,\beta+j)}(s; a, b, c - k). \qquad (6.2)$$

We now define the generalized moments of the weight function $\rho(s)$ by

$$c_{j,k} = \sum_{s=a}^{b-1} g_j(s) f_k(s) \rho(s) \nabla x_1(s). \qquad (6.3)$$

We set

$$B(\alpha, \beta, c) = \sum_{s=a}^{b-1} \rho^{(\alpha,\beta)}(s; a, b, c) \nabla x_1(s), \qquad (6.4)$$

These functions may be regarded as an extension of the beta function (see, for example, [2] and [18]), then in view of (6.2), we have

$$c_{jk} = B(\alpha + k, \beta + j, c - k). \qquad (6.5)$$

Representations of $u_n(s)$ and $v_n(s)$ in terms of the moments (6.3) then take the form

$$
u_n(s) = C \cdot \det \left(\begin{matrix} \int g_j f_k d\mu(s) \\ f_k \end{matrix} \right)_{\substack{j=0,1,\ldots,n-1 \\ k=0,1,\ldots,n}}
$$

$$
= \text{const.} \begin{vmatrix} c_{0,0} & c_{0,1} & \cdots & c_{0,n} \\ c_{1,0} & c_{1,1} & \cdots & c_{1,n} \\ \vdots & & & \\ c_{n-1,0} & c_{n-1,1} & \cdots & c_{n-1,n} \\ f_0(s) & f_1(s) & \cdots & f_n(s) \end{vmatrix}, \tag{6.6}
$$

and

$$
u_n(s) = D \cdot \det \left(\begin{matrix} \int g_k f_j d\mu(s) \\ g_k \end{matrix} \right)_{\substack{j=0,1,\ldots,n-1 \\ k=0,1,\ldots,n}}
$$

$$
= \text{const.} \begin{vmatrix} c_{0,0} & \cdots & c_{n,0} \\ c_{0,1} & \cdots & c_{n,1} \\ \vdots & & \\ c_{0,n-1} & \cdots & c_{n,n-1} \\ g_0(s) & \cdots & g_n(s) \end{vmatrix}, \tag{6.7}
$$

where C and D are some constants.

Wilson [22] suggested a method for direct calculation of these determinants. It is based on the following recurrence relations for the "beta function" defined in (6.4):

$$
B(\alpha+1,\beta,c-1) = \frac{\gamma(\alpha+1)\gamma(b-a+\alpha+\beta+1)}{\gamma(\alpha+\beta+2)\gamma(c-b)}
$$
$$
\frac{\delta(a+b+\alpha)\gamma(c-b+\beta+1)}{\gamma(c-a+\beta)\delta(c+a-1)} B(\alpha,\beta,c) \tag{6.8}
$$

and

$$
B(\alpha,\beta+1,c) = \frac{\gamma(\beta+1)\gamma(b-a+\alpha+\beta+1)}{\gamma(\alpha+\beta+2)\gamma(c-a+\alpha+\beta+2)}
$$
$$
\cdot \frac{\delta(a+b-\beta-2)\gamma(c-a+\beta+1)}{\gamma(c-b+\beta+2)\delta(b+c+\alpha+\beta+1)} B(\alpha,\beta,c), \tag{6.9}
$$

which follow from the Pearson equation for the summand in (6.4). Considerations of length prevent us from carrying out the rather lengthy calculations, so we refer the reader to Wilson's paper [22].

7. Concluding remarks

In this paper we have given a brief discussion of an extension of Hahn's approach in orthogonal polynomials to the set of biorthogonal rational functions given in (4.1) and (4.2). All classical orthogonal polynomials including the charm polynomials can be shown to follow as limiting cases of these functions. They deserve more indepth study than we have been able to do here. Examples of continuous biorthogonality that we have avoided in this work exist in Rahman's work [14-17]. In a subsequent paper we shall consider this question from the difference equation point of view. We shall also make an effort at classifying these rational functions according to their lattice types as well as the boundary conditions, both discrete and continuous.

References

[1] G.E. Andrews and R. Askey, *Classical orthogonal polynomials*, Polynômes orthogonaux et applications, Springer-Verlag, Berlin, Heidelberg, New York, 1985, pp. 36-62.

[2] R. Askey, *Beta integrals and q-extensions*, Proc. Ramanujan Centennial International Conference (eds. R. Balakrishnan, K.S. Padmanabhan and V. Thangaraj), Ramanujan Math. Soc., Annamalai University, Annamalainagar, 1988, pp. pp. 85-102.

[3] R. Askey and J.A. Wilson, *A set of orthogonal polynomials that generalize the Racah coefficients or 6-j symbols*, SIAM J. Math. Anal. **10** (1979), pp. 1008-1016.

[4] R. Askey and J.A. Wilson, *Some basic hypergeometric polynomials that generalize Jacobi polynomials*, Mem. Amer. Math. Soc. no. #319 (1985).

[5] N.M. Atakishiyev. G.I. Kuznetsov and S.K. Suslov, *A definition and a classification of the classical orthogonal polynomials*, in preparation.

[6] N.M. Atakishiyev. G.I. Kuznetsov and S.K. Suslov, *The Charm polynomials*, Proceedings of the Third International Symposium on Orthogonal Polynomials and Their Applications, (Berzinski, Gori and Ronveaux, eds.) J.C. Baltzer AG, Basel, Switzerland, 1991, pp. 15-16.

[7] P.L. Chebyshev, Complete collected works in 5 volumes, Izdat. Akad. Nauk. SSSR, Moscow, 1947-1957; *On interpolation of values of equidistants*, 1975, pp. 66-87. (in Russian)

[8] T.S. Chihara, *An Introduction to Orthogonal Polynomials*, Gordon and Breach, New York, 1978.

[9] G. Gasper and M. Rahman, *Basic Hypergeometric Series*, Cambridge University Press, 1990.

[10] W. Hahn, *Über Orthogonal Polynome, die q-Differenzengleichungen genügen*, Math. Nachr. **2** (1949), pp. 4-34.

[11] A.F. Nikiforov, S.K. Suslov and V.B. Uvarov, *Classical Orthogonal Polynomials of a Discrete Variable*, Nauka, Moscow, 1985 (in Russian); English translation, Springer-Verlag.

[12] A.F. Nikiforov and S.K. Suslov, Lett. Math. Phys. **11** no. 1 (1986), pp. 27-34, *Classical orthogonal polynomials of a discrete variable on nonuniform lattices*.

[13] M. Rahman, *Product and addition theorems for Hahn polynomials* (to appear).

[14] M. Rahman, *Families of biorthogonal rational functions in a discrete variable*, SIAM J. Math. Anal. **12** (1981), pp. 355-367.

[15] M. Rahman, *An integral representation of a $_{1}0\phi_9$ and continuous bi-orthogonal $_{1}0\phi_9$ rational functions*, Canad. J. Nath. **38** (1986), pp. 605-618.

[16] M. Rahman, *Biorthogonality of a system of rational functions with respect to a positive measure on* [−1, 1], SIAM J. Math. Anal. **22** (1991), pp. 1430–1441.

[17] M. Rahman, *Some extensions of Askey-Wilson's q-beta integral and the corresponding orthogonal systems*, Canad. Math. Bull. **31** (1988), pp. 467–476.

[18] M. Rahman and S.K. Suslov, *The Pearson equation and the beta integrals*, submitted.

[19] S.K. Suslov, *Classical orthogonality polynomials of a discrete variable, continuous orthogonality relations*, Lett. Math. Phys. **14** (1987), pp. 77–88.

[20] S.K. Suslov, *The theory of difference analogues of special functions of hypergeometric type*, Russian Math. Surveys **44** no. 2 (1989), pp. 227–278.

[21] J.A. Wilson, *Some hypergeometric orthogonal polynomials*, SIAM J. Math. Anal. **11** (1980), pp. 690–701.

[22] J.A. Wilson, *Orthogonal functions from Gram determinants*, SIAM J. Math. Anal. **22** (1991), pp. 1147–1155.

[23] J.A. Wilson, *Hypergeometric series recurrence relations some new orthogonal functions*, Ph.D. Thesis (1978), University of Wisconsin, Madison, Wisc..

*Dept. of Mathematics and Statistics   **Kurchatov Institute
Carleton University, Ottawa   of Atomic Energy, Moscow
Ontario, Canada K1S 5B6   123182, Russia.

A DIRECT PROOF
FOR TREFETHEN'S CONJECTURE

A.I. Aptekarev

Let $\mathcal{R}_n$ be the set of all rational functions of degree n and

$$L_n(R) = 2 \int_{\mathbf{T}} \frac{|R'_n(z)||dz|}{1 + |R_n(z)|^2}, \quad \mathbf{T} = \{z : |z| = 1\}$$

be the arc length of $R(\mathbf{T})$ on the Riemann sphere $\mathbf{S} = \{x \in \mathbb{R}^3 : |x| = 1\}$. At a meeting in Oberwolfach in February, 1991 L.N.Trefethen raised a problem of proving the following

THEOREM. *For all* $n \in \mathbb{N}$

$$L_n = \max_{R \in \mathcal{R}_n} L_n(R) = 2\pi n \tag{1}$$

This note presents a direct proof of this assertion. Another proof based on "Poincare's formula" was provided by E.Wegert. See [1], where one can find also an interesting discussion of the problem.

Proof. Let G_0 denote the group of linear fractional transformations of the unit disc and G_1 denote the group of linear fractional transformations generated by the rotations of $\mathbf{S}$.

LEMMA. *For any* $n \in \mathbb{N}$, $R_n \in \mathcal{R}_n$ *there exist* $g_0 \in G_0$ *and* $g_1 \in G_1$ *wich satisfy*

$$g_1 \circ R_n \circ g_0(z) = z R_{n-1}(z) \tag{2}$$

where $R_{n-1} \in \mathcal{R}_{n-1}$.

Proof. Let $\zeta \in \bar{\mathbb{C}}$ satisfy

$$R_n(\bar{\zeta}^{-1}) = -\overline{R_n(\zeta)}^{-1} \tag{3}$$

This equation always has solutions because both functions in $\bar{\zeta}$ on the left and right hand sides are rational. We can also suppose that $|\zeta| < 1$ since $\bar{\zeta}^{-1}$ and $\bar{\zeta}$ satisfy (3).

We choose now $g_0 \in G_0$ with $g(0) = \zeta$; then $g_0(\infty) = \bar{\zeta}^{-1}$. It follows by (2) that $R_n \circ g_0(0)$ and $R_n \circ g_0(\infty)$ are antipodal points on $\mathbf{S}$ and so there exists a rotation $g_1 \in G_1$ which maps these points to 0 and ∞ correspondingly. Hence $g_1 \circ R_n \circ g_0$ maps 0 to 0 and ∞ to ∞ and therefore satisfies (2).

Let R_n be an extremal function in (1). Then $L_n = L(R_n) = L(g_1 \circ R_n \circ g_0)$ since $g_0(\mathbf{T}) = \mathbf{T}$ and the arc length does not change under a rotation of $\mathbf{S}$. Now, it follows by lemma that

$$L_n = L(zR_{n-1}) = 2 \int_{\mathbf{T}} \frac{|R_{n-1} + zR'_{n-1}|}{1 + |R_{n-1}|^2} |dz| \leq L_{n-1} + 2\pi,$$

$n \in \mathbf{N}$, $L_0 = 0$. We have therefore $L_n \leq 2\pi n$. For $R_n = z^n$ we have equality and the proof is completed.

References

[1] E.Wegert and L.N.Trefethen, *The Arc Lengh of a Rational Function on the Riemann Sphere or From the Buffon Needle Problem to the Kreiss Matrix Theorem*, preprint.

Keldysh Institute Applied Mathematics
Moscow, Russia

A DIRECT PROOF
FOR TREFETHEN'S CONJECTURE

A.I. Aptekarev

Let $\mathcal{R}_n$ be the set of all rational functions of degree n and

$$L_n(R) = 2 \int_{\mathbf{T}} \frac{|R'_n(z)||dz|}{1 + |R_n(z)|^2}, \quad \mathbf{T} = \{z : |z| = 1\}$$

be the arc length of $R(\mathbf{T})$ on the Riemann sphere $\mathbf{S} = \{x \in \mathbf{R}^3 : |x| = 1\}$. At a meeting in Oberwolfach in February, 1991 L.N.Trefethen raised a problem of proving the following

THEOREM. *For all $n \in \mathbf{N}$*

$$L_n = \max_{R \in \mathcal{R}_n} L_n(R) = 2\pi n \tag{1}$$

This note presents a direct proof of this assertion. Another proof based on "Poincare's formula" was provided by E.Wegert. See [1], where one can find also an interesting discussion of the problem.

Proof. Let G_0 denote the group of linear fractional transformations of the unit disc and G_1 denote the group of linear fractional transformations generated by the rotations of $\mathbf{S}$.

LEMMA. *For any $n \in \mathbf{N}$, $R_n \in \mathcal{R}_n$ there exist $g_0 \in G_0$ and $g_1 \in G_1$ wich satisfy*

$$g_1 \circ R_n \circ g_0(z) = z R_{n-1}(z) \tag{2}$$

where $R_{n-1} \in \mathcal{R}_{n-1}$.

Proof. Let $\zeta \in \bar{\mathbf{C}}$ satisfy

$$R_n(\bar{\zeta}^{-1}) = -\overline{R_n(\zeta)}^{-1} \tag{3}$$

This equation always has solutions because both functions in $\bar{\zeta}$ on the left and right hand sides are rational. We can also suppose that $|\zeta| < 1$ since $\bar{\zeta}^{-1}$ and $\bar{\zeta}$ satisfy (3).

We choose now $g_0 \in G_0$ with $g(0) = \zeta$; then $g_0(\infty) = \bar{\zeta}^{-1}$. It follows by (2) that $R_n \circ g_0(0)$ and $R_n \circ g_0(\infty)$ are antipodal points on $\mathbf{S}$ and so there exists a rotation $g_1 \in G_1$ which maps these points to 0 and ∞ correspondingly. Hence $g_1 \circ R_n \circ g_0$ maps 0 to 0 and ∞ to ∞ and therefore satisfies (2).

This remark shows that harmonic vector fields (h.v.f.'s) are a natural 3-dimensional generalization of analytic functions of one complex variable (or, rather, that the latter are a 2-dimensional degeneration of h.v.f.'s). Usual problems of complex analysis (uniqueness theorems, maximum principles, normal families) become much more difficult if we try to solve them for h.v.f.'s. But *approximation properties* of complex analytic functions seem to be not so hard for generalization. In this article first steps in this direction are made. We prove multidimensional analogues of the Runge and Hartogs–Rosenthal theorems on rational approximation in $\mathbb{C}$. This is done not only for h.v.f.'s in $\mathbb{R}^3$, but in a more general setting, namely, for harmonic differential forms (of any degree) in $\mathbb{R}^n$.

Our predecessors (as far as we know) are Rao [1] and A.A.Shaginyan [2]. But their concern was approximation by harmonic *gradients*, which form a *proper* subclass of h.v.f.'s (a general h.v.f. coincides with a gradient only *locally*; see (2)). When reduced to $\mathbb{R}^2(=\mathbb{C})$, approximation by harmonic gradients becomes approximation by rational functions with a *single-valued primitive*; but our sample is classical rational approximation in $\mathbb{C}$ with no restrictions on approximants.

2. Elementary harmonic fields

Classical approximation theory in $\mathbb{C}$ created by Runge, Lavrentiev, Keldysh, Mergelyan and Vitushkin (see [3], [4]) is by now generalized in many respects and is nothing more, than a particular case of a large theory of approximation by solutions of systems of elliptic PDE's. The modern state of this theory is described in detail in [5]. Being highly developed, it doesn't apply, however, to (1) (and to the more general system

$$d\omega = 0, \quad \delta\omega = 0, \tag{3}$$

discussed below). The reason is this: symbols of systems, considered in [5] are *surjective* and solutions are representable as linear (integral) combinations of *point* singularities. E.g., general solutions of the *planar* Cauchy–Riemann system are more or less the same as "linear combinations" $\int \frac{d\mu(t)}{t-z}$ ($z \notin$ supp μ) of Cauchy kernels $z \mapsto (t-z)^{-1}$ with the only singularity at t. This is no longer the case with (1) (not to mention (3) for $n \geqslant 3$). There exist, of course, "elementary" solutions of (1) with one-point singularity, namely

$$\vec{V}(p) = \nabla_p \frac{1}{|p-a|} \quad (p \in \mathbb{R}^3 \setminus \{a\}).$$

But these "Coulomb fields" do not suffice to create (or to approximate) an arbitrary h.v.f. To see this consider a rectifiable path γ in $\mathbb{R}^3$:

$\gamma : [a,b] \to \mathbb{R}^3$, γ absolutely continuous, and $|\gamma'(t)| = 1$ a.e. in $[a,b]$.

APPROXIMATION PROPERTIES
OF HARMONIC VECTOR FIELDS
AND DIFFERENTIAL FORMS

V.P.Havin, A.Presa Sagué

This work is an attempt to find a multidimensional generalization of the planar theory of uniform rational approximation. Harmonic differential forms in $\mathbb{R}^n$ are considered as analogues of analytic functions in $\mathbb{C}$, whereas rational functions of a complex variable are replaced by the so-called Biot–Savard forms (with singularities on appropriate cycles instead of points). A Runge-like theorem is proved. Theorems by Hartogs–Rosenthal and Rao (on approximation by harmonic gradients in $\mathbb{R}^3$) are generalized to any dimension and any degree of forms.

1. Introduction.

Let U be an open subset of $\mathbb{R}^3$, $\vec{V} \to \mathbb{R}^3$ be a C^1-vector field. If

$$\operatorname{rot} \vec{V}(p) = 0, \quad \operatorname{div} \vec{V}(p) = 0 \quad (p \in U) \tag{1}$$

then $\vec{V}$ is called *harmonic*. Differential equations (1) mean that *locally* $\vec{V}$ is the gradient of a harmonic function:

$$\forall p \in U \quad \exists \delta > 0, \; h_p : \vec{V}|B(p,\delta) = \nabla h_p, \tag{2}$$

h_p is harmonic in $B(p,\delta)$ $(B(p,\delta) := \{q : |q - p| < \delta\})$.

Suppose f is a complex C^1-function, defined in a domain $O \subset \mathbb{C}$:

$$f = u - iv, \quad u, v \text{ real.}$$

Put $U := O \times \mathbb{R}(\subset \mathbb{R}^3)$, $\vec{V}(p) := (u(x,y), v(x,y), 0)$ $(p = (x,y,z) \in O)$. Then (1) is equivalent to

$$\frac{\partial u}{\partial y} - \frac{\partial v}{\partial x} = 0, \quad \frac{\partial u}{\partial x} + \frac{\partial v}{\partial y} = 0 \quad \text{in } O,$$

i.e. to the complex analyticity of f.

$$4\pi \vec{V}(p) = -\nabla_p \int_{\partial O} \frac{\langle \vec{V}(q), \vec{n}(q) \rangle}{|q - p|} dS(q) + \mathrm{rot}_p \int_{\partial O} \frac{\vec{n}(q) \times \vec{V}(q)}{|q - p|} dS(q), \quad (p \in O) \quad (6)$$

where O is a small neighbourhood of K with a nice boundary, dS is the area element on ∂O, $\langle \vec{a}, \vec{b} \rangle$ and $\vec{a} \times \vec{b}$ are, respectively, the scalar and the vector products, and $\vec{n}$ is the unit outer normal vector on ∂O (see, e.g., [6]). But this time integral sums don't do the job. Everything is O.K. with the first integral (taking integral sums, we obtain linear combinations of Coulomb fields), but not with the second: *a field of the form* $p \mapsto \mathrm{rot}_p \frac{\vec{a}}{|q-p|}$ *is not harmonic.* The harmonicity of the second summand in (6) stems from a delicate interaction of *non harmonic* fields, which are integrated over ∂O. This difficulty can be overcome by removing periods of $\vec{V}$ (by means of a suitable linear combination of Biot–Savard forms) and applying other integral formulae. But in the general case (see (3)) this is not so easy. Our proofs are non constructive and use duality and analysis of vector measures orthogonal to Biot–Savard differential forms.

In what follows we only state and briefly discuss the results.

4. Harmonic differential forms

Let us denote by $M_r (= M_{r,n})$ the set of all strictly increasing multiindices $\alpha = (\alpha_1, \ldots, \alpha_r)$, $\alpha_1 < \alpha_2 < \cdots < \alpha_r$, $\alpha_j \in \{1, 2, \ldots, n\}$. We write dx^α instead of $dx^{\alpha_1} \wedge \cdots \wedge dx^{\alpha_r}$, where x^l is the l-th coordinate in $\mathbb{R}^n$. We consider differential forms

$$\omega = \sum_{\alpha \in M_r} u_\alpha dx^\alpha$$

of degree r (r-forms), $0 \leqslant r \leqslant n$, u_α being (scalar) functions, defined on a set $E \subset \mathbb{R}^n$ (the domain of ω). We shall use standard notions and notation from [7]. In particular, $d\omega$ will denote the exterior differential and $\delta\omega$ the codifferential of ω. A form ω is called *harmonic*, if its domain is open, if it is C^1-smooth and if (3) holds. Any harmonic vector field $\vec{V} = (V_1, V_2, \ldots, V_n)$ in $\mathbb{R}^n$ can be identified with the harmonic 1-form $V_1 dx^1 + V_2 dx^2 + \cdots + V_n dx^n$.

5. Biot–Savard forms

Let γ be an $(n-r)$-dimensional cycle $((n-r)$-cycle) in $\mathbb{R}^n$ $(0 < r < n)$. Consider the following r-form:

$$\omega_{\gamma,r}(x) := \frac{1}{c_n} \sum_{\alpha \in M_r} \delta^{1 \ldots n}_{\alpha, \beta} \left(\int_\gamma \frac{d\xi^\beta}{|x - \xi|^{n-2}} \right) dx^\alpha, \quad (x \notin \Gamma := \mathrm{supp}\, \gamma)$$

($p \in \mathrm{supp}\, \gamma$ means that γ passes through p; $\delta^{1 \ldots n}_{\alpha_1 \ldots \alpha_r, \beta_1 \ldots \beta_{n-r}}$ denotes the Kronecker symbol; indices $\beta_1 < \beta_2 < \cdots < \beta_{n-r}$ fill the complement of α

in $\{1, \ldots, n\}$; we hope the reader won't be misled by the unpleasant, but unavoidable closeness of Kroneker's δ to δ as the codifferential; c_n is the $(n-1)$-dimensional volume of the sphere $\mathbb{S}^{n-1} \subset \mathbb{R}^n$).

Put
$$\omega_{\gamma,r} := \frac{1}{c_n} \frac{dx^1 \wedge \cdots \wedge dx^n}{|x - \xi|^{n-2}} \qquad (x \in \mathbb{R}^n \setminus \{\xi\})$$

for the "zero dimensional cycle" $\gamma := \{\xi\}$. Now everything is ready to define Biot–Savard forms BS^1_γ of the first kind. Namely, put

$$BS^1_{\gamma,r} := \delta \omega_{\gamma,r+1}, \tag{7}$$

where $0 < r < n$ and γ is an $(n-r+1)$-cycle. Biot–Savard forms BS^2_γ of the second kind are defined as follows:

$$BS^2_{\gamma,r} := *\delta \omega_{\gamma,n-r+1}, \tag{8}$$

where γ is an $(r-1)$-cycle, and $*$ is the usual operator, transforming p-forms into $(n-p)$-forms ($*dx^\alpha = \delta^{1\cdots n}_{\alpha,\beta} dx^\beta$, $\alpha \in M_p$). Both forms (7) and (8) are of degree r. It is not hard to prove (using the Stokes formula and the identity $d\delta + \delta d = \Delta$) that $BS^j_{\gamma,r}$ is harmonic outside of Γ.

Forms $BS^2_{\gamma,r}$ have no periods: if λ is an r-cycle, not intersecting Γ, then

$$\int_\lambda BS^2_{\gamma,r} = \int_\lambda *BS^1_{\gamma,n-r} = \pm \int_\lambda d(*\omega_{\gamma,n-r+1}) = 0$$

(recall that $\delta = \pm * d*$). Forms $BS^2_{\gamma,1}$ (where γ is a 0-cycle) can be identified with Coulomb vector fields.

BS-forms of the first kind may have non zero periods. Moreover,

$$\int_\lambda BS^1_{\gamma,r} = (-1)^{r+1} v(\lambda, \gamma), \tag{9}$$

where $v(\lambda, \gamma)$ is the linking coefficient of cycles λ and γ.

6. Runge's theorem

Suppose $\vec{V}$ is a h.v.f., defined in a neighbourhood O of a (3-dimensional, compact) torus $T \subset \mathbb{R}^3$. Let γ_1, γ_2 be 1-cycles in T, generating its fundamental group. Using (9), we can easily find numbers a_1, a_2, for which the field $\vec{W} := \vec{V} - a_1 BS_{\gamma_1} - a_2 BS_{\gamma_2}$ has no periods in O. Thus $\vec{W} = \nabla h$, where h is a function, harmonic in O. Consider an open torus O', $O \supset \text{Clos}\, O' \supset O' \supset K$, and represent h as a potential of a simple layer:

$$h(p) = \int_{\partial O'} \frac{k(q) dS(q)}{|q - p|}, \qquad (p \in O').$$

Taking integral sums for the last integral, we approximate $\vec{V}$ uniformly on T by linear combinations of Coulomb fields (BS-forms of the second kind) and of two Biot–Savard fields (BS-forms of the first kind).

This simple scheme can be used to state and prove Runge's theorem for h.v.f.'s in $\mathbb{R}^n$ (or, what is the same, for harmonic 1-forms). But the proof of the more general assertion concerning harmonic forms of any degree (see below) is not as direct.

Let $(\gamma_i^1)_{i=1}^l$, $(\gamma_i^2)_{i=1}^l$ be two families of $(n-r-1)$- and $(r-1)$-cycles (respectively), and $(a_i^1)_{i=1}^l$, $(a_i^2)_{i=1}^l$ two families of real numbers. Put

$$BS_{c_j}^j := \sum_{i=1}^l a_i^j BS_{\gamma_i^j}^j, \quad c_j := \sum_{i=1}^l a_i^j \gamma_i^j, \quad j = 1, 2, \tag{10}$$

(Biot–Savard forms, generated by chains c_1 and c_2). Let us call an r-form representable as $BS_{c_1}^1 + BS_{c_2}^2 (=: BS_{c_1,c_2})$ a *quasirational form generated by chains c_1 and c_2*. If $\operatorname{supp}\gamma_i^1 \cup \operatorname{supp}\gamma_i^2 \subset E \subset \mathbb{R}^n$, $i = 1,\ldots,l$, then we say that c_1 and c_2 lie in E; if γ_i^1, γ_i^2 are homologous to zero in E, then we say that c_1 and c_2 are homologous to zero in E. We call BS_{c_1,c_2} the form of the first (second) kind if $a_i^2 \equiv 0$ (resp. $a_i^1 \equiv 0$).

Let K be a compact subset of $\mathbb{R}^n$, U an open subset of $\mathbb{R}^n \setminus K$ *intersecting every component of $\mathbb{R}^n \setminus K$*. Suppose u is an r-form harmonic in a neighbourhood O of K.

THEOREM. *Suppose $0 < r < n$. There exist chains c_1, c_2 (with cycles γ_i depending on K and O only), lying in $\mathbb{R}^n \setminus K$ and such that the form $u - BS_{c_1,c_2}$ can be approximated uniformly on K by quasirational forms u_j, generated by chains, lying in U and homologous to zero in U. Moreover, u_j can be taken of the same kind (all of the first kind or all of the second).*

7. Approximation on small sets

Let $C_r(K)$ denote the set of all r-forms continuous on K, a compact subset of $\mathbb{R}^n$, and let $\mathcal{H}_r(K)$ be the set of all restrictions to K of r-forms, harmonic in a neighbourhood of K (depending on the form). The set $C_r(K)$ becomes a Banach space in a natural way, and we may ask, for which K's

$$C_r(K) = \operatorname{Clos} \mathcal{H}_r(K) \tag{11}$$

If $n = 2$, $r = 1$, then this property coincides with uniform density of rational functions of one complex variable in $C(K)$, the space of all complex functions, continuous on K. In this case the problem was solved by Vitushkin in the sixties. But here we stay on the level of the thirties, when the Hartogs–Rosenthal theorem was proved, stating that (11) *holds if $n = 2$, $r = 1$ and*

$m_2(K) = 0$ (m_2 is the 2-dimensional Lebesgue measure). To generalize this theorem it will be convenient to change our point of view and replace $\mathbb{R}^n$ by an arbitrary n-dimensional oriented euclidean space E^n. Operators $d, *, \delta$ have an invariant meaning, not depending on the choice of the orthonormal basis; the same is true for the definition of harmonic forms and BS-forms.

Let K be a set in $E^n, 1 \leqslant r < n$. Let us call K r-invisible, if there exists an orthonormal basis in E^n such that all orthogonal projections of K onto coordinate planes of dimension $n - r$ are of Lebesgue $(n - r)$-measure zero. For instance, if K is 1-invisible in E_3, then for a triple e_1, e_2, e_3 of unit mutually orthogonal vectors any beam of rays, parallel to e_j, does not notice K "almost surely", because almost all rays of the beam avoid K. The property, described in the definition, *does* depend on the basis. Any set of $(n - r)$-dimensional Hausdorff measure zero is r-invisible, but not vice-versa.

Suppose $n \geqslant 3$ and K is a compact subset of E^n, $n \geqslant 3$, $1 \leqslant r < n$. Put $r' := max(r - 1, n - r - 1)$

THEOREM. *If K is r'-invisible, then (11) holds.*

If $n = 2$, $r = 1$, then $r' = 0$. It is natural to call 0-invisible a subset of E^k of zero n-dimensional measure, and the proof shows that, indeed, for $n = 2$, $r = 1$ we come to the Hartogs–Rosenthal theorem.

If $n = 3$, $r = 1$, then (11) means that any vector field, continuous on K, is a uniform limit of fields, harmonic near K. Our theorem shows that this is the case, whenever K is 1-invisible (or of 2-dimensional Hausdorff measure zero). This result is sharp (be it in a very rough sense), as can be seen from the following example. Let K be a two-dimensional square in $\mathbb{R}^3$ (say, $K := \{(x, y, z) : z = 0, |x| \leqslant 1, |y| \leqslant 1\}$ so that K is not 1-invisible). Then $C_1(K) \neq \text{Clos}\,\mathcal{H}_1(K)$, because K contains 1-cycles. But any 1-cycle in K is an obstacle to approximation by gradients (whose period along any 1-cycle is zero), whereas any field, harmonic near K, *is* a gradient.

8. Approximation by closed and coclosed harmonic forms

To escape from the above obstacle (possible presence of cycles in K), let us change (11), replacing $C_r(K)$ by its appropriate subclass. Namely, let $C_r^d(K)$ (resp. $C_r^\delta(K)$) be the closure in $C_r(K)$ of all forms $\omega | K$, where ω runs through the class of all r-forms, smooth and closed (resp. coclosed) in a neighbourhood of K. We shall also consider classes $C_r^D(K)$ and $C_r^\Delta(K)$, obtained in the same way from smooth r-forms ω, exact and coexact in a neighbourhood of K. So, instead of (11), we are now interested in the following properties:

$$(a) \ \text{Clos}\,\mathcal{H}_r(K) = C_r^d(K), \ (b) \ \text{Clos}\,\mathcal{H}_r(K) = C_r^\delta(K),$$
$$(c) \ \text{Clos}\,\mathcal{H}_r(K) \supset C_r^D(K), \ (d) \ \text{Clos}\,\mathcal{H}_r(K) \supset C_r^\Delta(K) \tag{12}$$

It is easier for K to satisfy one of these conditions, than (11). It is not hard to show, that $(a) \Leftrightarrow (c)$, $(b) \Leftrightarrow (d)$. Rao proved (in [1]) that if $r = 1$, then $m_n(K) = 0$ implies (c).

THEOREM. *Suppose K is a compact subset of E^n, $1 \leqslant r < \frac{n}{2}$ (or $\frac{n}{2} < r < n$). If K is $(r-1)$-invisible (resp. $(n-r-1)$-invisible), then (12a) (resp. (12b)) holds.*

We have here $r - 1 < n - r - 1$ (or $n - r - 1 < r - 1$), and invisibility conditions of this theorem are weaker, than those of the preceding theorem, when our aim was the stronger property (11).

9. A remark on approximation in $L^p(K)$

Consider the space $L_r^p(K)$ of all r-forms, whose coefficients belong to $L^p(K, m_n)$ ($K \subset E^n$ is compact, m_n is the n-dimensional Lebesgue measure in E^n).

THEOREM. *Suppose $1 < p < \frac{n}{(n-1)}$, $0 < r < n$. Put $r' := \max(n - r - 1, r - 1)$. If there is an orthonormal basis in E^n such that all orthogonal projections of K onto $(n - r')$-dimensional coordinate planes are nowhere dense, then*

$$\text{Clos}_{L^p} \mathcal{H}_r(K) = L_r^p(K). \tag{13}$$

This result generalizes a theorem of Sinanian [8] on the complex rational approximation in L^p for $1 < p < 2$. The theorem is sharp (in a sense). Put $n = 3$, $r = 1$, $K := k \times c$, where $k := \{(x,y) \in \mathbb{R}^2 : x^2 + y^2 \leqslant 1\}$, $c \subset \mathbb{R}$ is a Kantor set of positive length. Any orthogonal projection of K onto a two-dimensional plane has interior points, and it is easy to see that (13) isn't true.

References

[1] Rao N.V., *Approximation by gradients*, Journal of App. Th., **12** (1974), pp. 52–60.

[2] Shaginyan A.A., *On potential approximation of vector fields*, Lecture Notes in Math., **1275** (1987), pp. 272–279.

[3] Gamelin T.W., *Uniform algebras*, Prentice Hall, NJ, 1969.

[4] Смирнов В.И., Лебедев Н.А., *Конструктивная теория функций комплексного переменного*, Наука, М., 1964.

[5] Тарханов Н.Н., *Аппроксимация на компактах решениями систем с сюрвективным символом*, 65 стр. (препринт), Красноярск, 1989.

[6] Джураев А.Д., *Метод сингулярных интегральных уравнений*, М., Наука, 1987.

[7] G.De Rham, *Variétés différentiables*, Paris, Hermann, 1955.

[8] Синанян С.О., *Аппроксимация аналитическими функциями и полиномами в среднем по площади*, Матем. сборник, **69** (1966), с. 546–578.

Sankt-Petersburg
State University
199155 Sankt-Petersburg
Russia

A PROBLEM OF AXLER AND SHIELDS ON NONTANGENTIAL LIMITS AND MAXIMAL IDEAL SPACE OF SOME PSEUDOANALYTIC ALGEBRAS

Oleg V. Ivanov

1. Introduction

In [1] Sheldon Axler and Allen Shields proved that every continuous function (with values in $\mathbb{C} \cup \{\infty\}$) on the open unit disk that admits a continuous extension to the Shilov boundary of H^∞ must have a nontangential limit at almost every point of the unit circle. A natural question arises if the converse to Axler-Shields Theorem holds true. In [2] we proved that this question has an affirmative answer. More precisely, if a continuous function on the open unit disk has a nontangential limit at almost every point of the unit circle then this function has a continuous extention to the Shilov boundary of H^∞.

Recall that $M(H^\infty)$ denotes the maximal ideal space of H^∞. Thus $M(H^\infty)$ is the set of multiplicative linear functionals from H^∞ to the field of complex numbers $\mathbb{C}$. Endowed with the usual topology, the weak-star topology which $M(H^\infty)$ inherits as a subset of the dual of H^∞, $M(H^\infty)$ becomes a compat Hausdorff space.

We can naturally embed the unit disk $\mathbb{D}$ to $M(H^\infty)$ by indentifying each point of the disk with the multiplicative linear functional of point evalution. The topology which $\mathbb{D}$ inherits as a subset of $M(H^\infty)$ coincides with the usual topology on $\mathbb{D}$. It is well-known that $\mathbb{D}$ is dense in $M(H^\infty)$. Every function from H^∞ regarded as a function from $\mathbb{D}$ to $\mathbb{C}$ can be extended, via the Gelfand transform, to a continuous compex-valued function defined on $M(H^\infty)$. Recall that [3]

$$C(M((H^\infty)) = \mathrm{alg}(H^\infty, \overline{H^\infty}).$$

If $f \in C(M(H^\infty))$, then $f|_D$ has a nontangential limit at almost every point of $\partial \mathbb{D}$.

Carl Sundberg [4] proved that every function in **BMO** extends to a continuous function from $M(H^\infty)$ to the Rieman sphere $\mathbb{C} \cup \{\infty\}$. He also [1,sec.5] gave an example of a function in the Bloch space that cannot be extended to a continuous function from $M(H^\infty)$ to $\mathbb{C} \cup \{\infty\}$.

However, Leon Brown and P.M.Gauthier [5] proved that every normal function (in particulary, every Bloch function) can be extended to a continuous function (with values in $\mathbb{C} \cup \{\infty\}$) defined on the union of all nontrivial Gleason parts of $M(H^\infty)$.

DEFINITION. Let f be a continuous function $f : \mathbb{D} \to \overline{\mathbb{C}}$. We say that f satisfies the Fatou condition $f \in (F)$ notationally $f \in (F)$ if $f|D$ has a nontangential limit at almost every point of $\partial \mathbb{D}$.

Recall that $X = M(L^\infty) \subset M(H^\infty)\backslash\mathbb{D}$ is the Shilov boundary of H^∞.

THEOREM (Axler-Shields). *Let f be a continuous function on $\mathbb{D} \cup X$. Then $f \in (F)$.*

Our main result is the following.

THEOREM 1. *Let $f \in (F)$. Then $f|_D$ has a continuous extension*

$$\tilde{f} : \mathbb{D} \dot\cup X \to \overline{\mathbb{C}}.$$

The following brilliant fact was proved by Gamelin [6,p.31]

LEMMA 1. *Let S be any subset of $\mathbb{D}$. Then $[S]_M \cap M(L^\infty) \neq \emptyset$ if and only if $m(F(S)) > 0$, where m is the Lebesque measure and $F(S)$ is a set nontangential cluster points of S.*

If there exists an angle Γ and a sequence $\{z_n\}$ poits of $S \cup \Gamma$ satisfying $\lim |z_n - z_0| = 0$, then a point $z_0 \in S^- \cap \mathbb{T}$ is called a nontangential cluster point of S.

We are now in a position to prove Theorem 1.

STEP 1. Assume that $f \in (F)$ and f is a bounded function. Let F be a complex harmonic function such that $\tilde{F}|_\mathbb{T} = \tilde{f}|_\mathbb{T}$, where $\mathbb{T} = \{z : |z| = 1\}$. Then nontangential limits of $\phi = F - f$ are almost everywhere zero. Assuming that ϕ has no zero extension to X we see that there exists a point $x \in X$ such that $\limsup\limits_{z \to x} |\phi(z)| = \alpha > 0$. Let $U \overset{\text{def}}{=} \{z : |\phi(z)| > \alpha/2\}$. It is easy to see that $x \in [U]_M \cap X \neq \emptyset$. Then by Lemma 1 $F(U) > 0$ which contradicts to the fact that nontangential limits of ϕ vanish almost everywhere.

STEP 2. To complete the proof we consider the final case where f is a continuous function on $\mathbb{D}$ taking values in $\mathbb{C} \cup \{\infty\}$. Let $\mathbb{P} : \overline{\mathbb{C}} \to \mathbb{C}$ be a continuous function such that for every $w \in \mathbb{C}$, $\mathbb{P}^{-1}(w)$ is a set consisting of one or two points [1]. Let $Cl(f, t) = \cap_{O_\alpha} [f(O_\alpha(t) \cap \mathbb{D})]^-$ be a cluster set of f at a point $t \in X$, where $\{O_\alpha\}$ is a basis of neighbourhoods at t in $M(H^\infty)$. It is easy to see that $Cl(f, t)$ consists of one or two points. Thus Lemma 1 implies that $Cl(f, t)$ consists of only one point. This means that for any point $t \in X$ there exists a limit.

Let $\mathcal{H}_{H^\infty}$ and $\mathcal{H}_{L^\infty}$ be Banach algebras of bounded continuous functions having nontangential limits on $\mathbb{T}$ in $H^\infty(\mathbb{T})$ and $L^\infty(\mathbb{T})$ respectively, $\|f\| = \sup |f(z)|$. It was proved in [7-9] that

$$M(\mathcal{H}_{H^\infty}) = M(H^\infty) \cup M(\mathcal{L}_{H^\infty}) \text{ and } M(H^\infty) \cap M(\mathcal{H}_{H^\infty}) = M(L^\infty).$$

The maximal ideal space $M(\mathcal{H}_{H^\infty})$ of $\mathcal{H}_{H^\infty}$ is a union of two compactifications $M(H^\infty)$ and $M(\mathcal{H}_{L^\infty})$, which is spliced along the compact $X = M(L^\infty)$. Let $\pi : \beta\mathbb{D} \to \mathcal{H}_{L^\infty}$ be a continuous mapping $\pi(z) = z$, $z \in \mathbb{D}$.

THEOREM 2, [10]. $M(\mathcal{H}_{L^\infty}) \backslash (\mathbb{D} \cup M(L^\infty)) = \beta\mathbb{D} \backslash (\mathbb{D} \cup \pi^{-1}(M(L^\infty)))$.

COROLLARY. *Any bounded continuous function $f : \mathbb{D} \to \mathbb{C}$ has a continuous extension to $\mathcal{H}_{L^\infty} \backslash (\mathbb{D} \cup M(L^\infty))$.*

Acknowledgements. I am deeply indebted to Professor Hakan Hedenmalm for many valuable discussions during his visit to Donetsk. Also I thank Professors Alexander Volberg and Vadim Tolokonnikov for having simplified some parts of the proof of Theorem on the maximal ideal space $M(\mathcal{H}_{H^\infty})$.

References

[1] Axler A., Shields A., *Extensions of harmonic and analytic functions*, Pacif. J. Math. **145** no. 1 (1990), pp. 1–15.

[2] Ivanov O.V., *Fatou theorem on nontangential limits and questions of extension on the ideal boundary*, Zap. LOMI **190** no. 19 (1991), pp. 101–109.

[3] Hoffman K., *Bounded analytical functions and Gleason parts*, Ann. Math. **86, 1967** no. 1, pp. 74–111.

[4] Sundberg C., *Truncations of BMO functions*, Indiana Univ. Math. J. **33** (1984), pp. 749–771.

[5] Brown L., Gauthier P. M., *Behavior of normal meromorphic functions on the maximal ideal space of H^∞*, Mich. Math. J. **18** (1971), pp. 365–371.

[6] Gamelin T.W., *Lecture on $H^\infty(\mathbb{D})$*, Universidad national de la Plata (1972), pp. 1–99.

[7] Ivanov O.V., *Generalize analytic functions and a Two-Leaf Corona theorem*, Dokl. Akad. Nauk Ukrain. SSR ser.A no. 4 (1989), pp. 10–11.

[8] Ivanov O.V., *Generalized analytic functions and analytic subalgebras*, Ukrain. Math. J. **42** no. 5 (1990), pp. 616–620.

[9] Ivanov O.V., *Generalize Douglas algebras and Corona theorem*, Subirean Math. J. **32** no. 1 (1991), pp. 37–42.

[10] Ivanov O.V., *Nontangential limits and Shilov boundary of algebra H^∞*, Dokl. Akad. Nauk. Ukrain. SSR ser.A no. 7 (1991).

Inst. of Applied Math. and Mech.
Ukrainian Academy of Sc.
Donetsk Ukraina

DEGREE OF APPROXIMATION
OF ANALYTIC FUNCTIONS
BY "NEAR THE BEST"
POLYNOMIAL APPROXIMANTS

V.V. Maimeskul

1. Introduction

Let K – be a compact subset of the complex plane $\mathbb{C}$, with connected complement Ω, $A(K)$ be the class of functions that are continuous on K and analytic in the interior K° of K. The rate of the best uniform approximation of a function $f \in A(K)$ by polynomials of degree not exceeding n is denoted by $E_n(f) = E_n(f, K)$:

$$E_n(f) := \inf\{\|f - P_n\|_{C(K)} : \deg P_n \le n\};$$

We denote by $P_n(f, z)$ the polynomial minimizing $\|f - P_n\|$. It was shown in [1] that $|f(z) - P_n(f, z)|$ do not decrease faster then $E_n(f)$ at interior points of K (for all f in $A(K)$ except of a subset of the first category). At the same time, under some assumption on the geometry of K it was proved that there exist such "near the best" polynomial approximants f on K which approximate f more rapidly at compact subsets of K°.

We prove analogous result for more general class of compact sets.

2. Main results

THEOREM 1. *Let $K = \cup_{i=1}^s K_i$ where K_i, $i = \overline{1,s}$ are compact sets with connected interiors and complements, and $K_i \cap K_j = $ for $i \ne j$; $f \in A(K)$. Then there exist polynomials p_n, $n = 1, 2, \ldots, \deg p_n \le n$, such that*

$$\varlimsup_{n \to \infty} \frac{\|f - p_n\|_{C(K)}}{E_n(f)} \le 2 \tag{1}$$

and

$$\lim_{n \to \infty} \frac{f(z) - p_n(z)}{E_n(f)} = 0 \tag{2}$$

uniformly on compact subsets of K°.

The above mentioned result by E.Saff and V.Totik is a special case of Theorem 1 for $s = 1$. The constant 2 in (1) is the best possible ([1], Theorem 3). Our next result shows that "near the best" polynomial approximants can converge more rapidly than $P_n(f, z)$ on compact subsets of K° for arbitrary compact K (Ω is assumed to be connected).

THEOREM 2. *There exists a constant $C > 0$ such that for every compact K and $f \in A(K)$ there exist polynomials p_n $n = 1, 2, \ldots, \deg p_n \leq n$, such that*

$$\|f - p_n\|_{C(K)} \leq C E_n(f) \tag{3}$$

and

$$\lim_{n \to \infty} \frac{f(z) - p_n(z)}{E_n(f)} = 0 \tag{4}$$

uniformly on compact subsets of K°

Unlike to Theorem 1 we do not know the precise value of C in (3), but it can be easily estimated.

In the case when $K = \bar{G}$ is the closer of a domain with analytic boundary the statement (4) of Theorem 2 can be improved [1]:

$$\varlimsup_{n \to \infty} \left(\frac{|f(z) - p_n(z)|}{E_n(f)} \right)^{1/n} > 1. \tag{5}$$

As N.Shirokov has informed us (private communication) this result fails if the boundary ∂G is not analytic. In this connection E.Saff posed a problem on existence of possible analogues of (5) for arbitrary compacts. The following theorem provides a step in this direction.

THEOREM 3. *For an arbitrary compact K with connected complement there exists a monotonically increasing function $\varphi(x)$, $\varphi(\infty) = \infty$, such that for every f in $A(K)$ there are polynomials p_n, $n = 1, 2, \ldots, \deg p_n \leq n$, satisfying*

$$\|f - p_n\|_{C(K)} \leq C E_n(f) \quad \text{and} \quad \lim_{n \to \infty} \left(\frac{f(z) - p_n(z)}{E_n(f)} \varphi(n) \right) = 0$$

uniformly on compact subsets of K°, where C is an absolute constant.

3. Preliminaries

To prove our results we need a few auxiliary lemmas. The following lemma is a simple corollary of a general result by Walsh (see, for example, [2], Ch. IV, Theorem 5).

LEMMA 1. *Let $K = K_1 \cup K_2$ where K_i $i = 1, 2$, are the compacts with connected complements and $K_1 \cap K_2 = \emptyset$; $g(z)$ be the characteristic function of K_1. There exist constants $\rho = \rho(K) < 1$, $C_1 = C_1(K)$ and a sequence $\{q_l(z)\}_{l=1}^\infty$ of polynomials q_l of degree $\leq l$ such that*

$$\|g - q_l\|_{C(K)} < C_1 \rho^l.$$

LEMMA 2. *Let K be a compact with connected complement. Then there exists a compact $\mathfrak{M} \supset K$ such that*

a) *the complement $\tilde{\Omega} = C \backslash \mathfrak{M}$ is simply connected;*
b) $\mathfrak{M}^\circ = K^\circ$.

Given a compact set K we denote by $K_t = \cup_{z \in K}\{\zeta \in C : |\zeta - z| < t\}$ a t-neighborhood of K.

LEMMA 3. *There exists an absolute constant $C_2 > 0$ such that for any compact K with connected complement, any compact subset $E \subset K^\circ$ and $\epsilon > 0$ there exists $t = t(K, E, \epsilon)$ such that for every f in $A(K)$ there exists a function $R(z)$ analytic on K_t such that:*

a) $\|R\|_{C(K_t)} \leq C_2 \|f\|_{C(K)}$;
b) $\|f - R\|_{C(E)} \leq \epsilon$.

Proof of Lemma 3 is analogous to the proof of the corresponding result by Mergelyan (see, for example, [3], Ch.1, §7.4, Theorem 6).

LEMMA 4. *Let K be a compact with simply connected complement. Then for every $t > 0$, and every $\epsilon > 0$ there exists $n = n(K, t, \epsilon)$ such that for every R in $A(K_t)$ with $\|R\|_{C(K_t)} \leq 1$ there exists a polynomial $T_n(z)$ of degree not exceeding n such that*

$$\|R - T_n\|_{C(K)} < \epsilon.$$

Proof of Lemma 4 uses Prypik's method of the proof of the Bernstein-Walsh theorem ([4, Ch. IX, §9, Theorem 4) and some facts of geometric function theory.

4. Schemes of proofs of Theorems 1 and 2

Proof of Theorem 1. We show here what changes must be made in proof of Theorem 2 of [1] to pass from case $s = 1$ to case $s = 2$. For arbitrary finite s, the proof is analogous.

Let $R_{k,n}^{(i)}(f, z)$ be a good enough polynomial approximants of the degree at most $n^{(i)}(k)$, $k \geq 5$, constructed on K_i, $i = 1, 2$ for the function

$$Q_n(f, z) = \frac{f(z) - P_n(z)}{E_n(f)}, \tag{5}$$

the same as in the proof of Theorem 2 [1]. By Bernstein-Walsh's lemma (see, for example, [2], Ch. IV, §4.6) we have also some estimates for sup-norms of these polynomials on the whole K.

Consider an expression

$$T_{k,n}(f, z) = R_{k,n}^{(1)}(f, z)q_l(z) + R_{k,n}^{(2)}(f, z)(1 - q_l(z)),$$

where $q_l, l = 1, 2, \ldots$, are polynomials from Lemma 1. Choosing $l = l(k)$, we can achieve needed estimates for the approximation of $Q_n(f, z)$ by these polynomials on K. Degree of polynomials $T_{k,n}(f, z)$ is at most $n(k) + l(k) = n'(k)$. It is obvious, that the sequence $n'(k)$ can be choosed increasing. Thus, by setting in the same way as in [1],

$$p_{n0}(z) := P_n(f, z) + T_{k,n}(f, z)E_n(f)$$

for $n'(k) \leq n < n'(k+1)$, we get the required sequence of polynomials.

Proof of Theorem 2. First using Lemma 2 we construct a continuum $\mathfrak{M}$. Let $Q_n(f, z)$ be defined by (5). We extend these functions continuously to the whole plane and obtain continuous functions with compact support and the same norm. Then for each n $Q_n(f, z) \in A(\mathfrak{M})$ and it's sup-norm is at most 1.

Let $\{d_m\}_1^\infty$, $\{\epsilon_m\}_1^\infty$ – be arbitrary monotonically decreasing sequences, which converge to zero, $K_m = \{z \in K : \text{dist}(z, \Omega) > d_m\}$, $\{t_m\}_1^\infty$ – the numerical sequence defined by Lemma 3 for $E = K_m$ and $\epsilon = \epsilon_m$. Choosing $\epsilon = \epsilon_m$ and $t = t_m$ in Lemma 4, it is possible to define the sequence $n(K, m)$ to be increasing. The consecutive application of Lemmas 3 and 4 to functions Q_n under $n(K, m) \leq n < n(K, m + 1)$ ensure the existence of polynomials $T_{m,n}(z)$ such that

$$\|T_{m,n}\|_{C(K)} \leq C_3;$$

$$\|Q_n - T_{m,n}\|_{C(K_m)} \leq \epsilon_m$$

hold true with some absolute constant C_3. Then the polynomials

$$p_n(z) := P_n(f, z) + T_{m,n}(z)E_n(f)$$

form a sequence to be found because K_m is an increasing sequence of sets that converges to K°.

The proof of Theorem 3 is analogous.

References

[1] E.B.Saff and V.Totik, *Behavior of polynomials of best uniform approximation*, Trans. Amer. Math. Soc. **316** no. 2 (1989), pp. 567–593.

[2] J.L.Walsh, *Interpolation and approximation by rational functions in the complex domain*, Publ. by AMS, 1960, Second Ed..

[3] V.I.Smirnov and N.A.Lebedev, *Constructive theory of functions of complex variable*, Nauka, M.-L., 1964. (In Russian)

[4] V.K.Dzjadyk, *Introduction in the theory of the uniform approximation of functions by polynomials*, Nauka, M., 1977. (In Russian)

Institute of Applied Mathematics and Mechanics
Ukrainian Academy of Sciences
Donetsk USSR

EXTREMAL PROBLEMS
FOR BLASCHKE PRODUCTS AND WIDTHS

O.G. Parfenov

1. Introduction

The paper is devoted to calculation and estimation of n-widths of classes of analytic functions. Recall the main definitions. Let X, Y be Banach spaces, T be a compact linear operator from X to Y. Define the numbers $d^n(T)$, $d_n(T)$, $l_n(T)$ as follows:

$$d^n(T) = \inf_{L_{-n}} \|T|_{L_{-n}}\|_{X \to Y},$$

$$d_n(T) = \inf_{L_n} \sup_{x \in BX} \inf_{y \in L_n} \|Tx - y\|_Y,$$

$$l_n(T) = \inf_{T_n} \|T - T_n\|_{X \to Y}.$$

Here L_{-n} is an arbitrary subspace of X codim$L = n$; L_n is an arbitrary subspace of Y, dim$L_n = n$; BX is the unit ball of X, T_n is an arbitrary linear operator from X to Y of finit rank n. The numbers $d^n(T)$, $d_n(T)$, $l_n(T)$ are called Gelfand n-widths, Kolmogorov n-widths, and linear n-widths.

Now let $\mathbb{D}$ be the unit disc in the complex plane $\mathbb{C}$ and $\mathbb{T}$ be the unit circle. As usual H^p denotes the Hardy space in $\mathbb{D}$. Consider the embedding operator

$$J: \ H^p \to \dot{\mathcal{L}}_q(\mu), \tag{1}$$

where μ is a finite Borel measure in $\mathbb{D}$. The n-widths of the J for $p = \infty$ were calculated by means of Blaschke products in [1]. Recently Fisher and Stessin have established a more general result.

THEOREM 1 (Fisher and Stessin, preprint 1990, Evanston). *Let* $1 \leq q \leq p \leq \infty$. *Then for every* n *we have*

$$d^n(J) = d_n(J) = l_n(T) = \inf_{B_n} \sup_{f \in BH^p} \|B_n f\|_{\mathcal{L}_q(\mu)}, \tag{2}$$

Here B_n is an arbitrary Blaschke product of degree n.

Denote the support of μ by supp μ. If supp μ is compact (i.e. lies strictly inside $\mathbb{D}$) then it is easy to prove that for any p, q

$$d^n(J) \asymp d_n(J) \asymp l_n(T) \asymp \inf \|B_n\|_{\mathcal{L}_q(\mu)}. \tag{3}$$

(As usual the notation $a_n \asymp b_n$ means that there exist two constants c_1, c_2 such that $c_1 a_n \leq b_n \leq c_2 a_n$).

The assumption about compactness of supp μ is essential. Indeed let $p = q = 2$ and $d\mu$ be the Lebesque measure dx on $[0, 1]$. Then the operator J is bounded and noncompact. This can be easily proved using the Carleson embedding theorem. On the other hand clearly

$$\inf_{B_n} \|B_n\|_{\mathcal{L}_2([0,1])} \xrightarrow[n \to \infty]{} 0. \tag{4}$$

The precise rate of decrease of sequence (4) was established in [2].

In case $p \geq q$ and supp $\mu = T_r$, $T_r = \{z : |z| = r\}$, where $r < 1$, explicit formulae for n-widths were given by the author [3].

The present paper is devoted to the case when supp μ is noncompact in $\mathbb{D}$. Denote by dS the area measure in $\mathbb{D}$.

THEOREM 2. *Suppose that $p \geq q$ and $d\mu = \phi dS$, for some $\phi \in C(\bar{\mathbb{D}})$. Then*

$$d^n(J) \sim \left(\frac{1}{2\pi} \int_0^{2\pi} \left(\phi(e^{it}) \right)^{\frac{p}{2p-q}} dt \right)^{\frac{2p-q}{pq}} (nq)^{-1/q}. \tag{5}$$

(Here $a_n \sim b_n$ means that $\lim_{n \to \infty} a_n b_n^{-1} = 1$).

This result is nontrivial only when the set supp $\mu \cap \mathbb{T}$ has positive linear measure. It is interesting to study what happens in the case $m(\text{supp }\mu \cap \mathbb{T}) = 0$. We state two results of this sort without proof. Define the set

$$\Delta_r = \{z \mid |z - (1 - r)| < r\}, \quad 0 < r < 1.$$

THEOREM 3. *Let J be the embedding operator from H^∞ into $\mathcal{L}_q(\Delta_r, dS)$. Then*

$$d^n(J) \asymp n^{-3/q}.$$

Let $G_\lambda = \{z \mid |\arg(1 - z)| < \lambda\} \cap \Delta_r$, $0 < \lambda < \frac{\pi}{2}$.

THEOREM 4. *Let J be the embedding operator from H^∞ into $\mathcal{L}_q(G_\lambda, dS)$. Then*

$$\ln d^n(J) \asymp \sqrt{n}.$$

Acknowledgments. Part of this paper was written in Linköping University where the author was a visitor in October 1991. The author thanks the staff of Linköping University (especially L.Hedberg and A.Laptev) for wonderful hospitality.

2. Proof of Theorem 2

LEMMA 1. *Let J be the embedding operator of H^p into $\mathcal{L}_q(\mathbb{D}, dS)$ and $p \geq q$. Then*

$$d_n(J_0) = (nq + 2)^{-1/q}, \qquad n = 0, 1, \ldots. \tag{6}$$

Proof. By Theorem 1

$$d_n(J_0) = \inf_{B_n} \sup_{f \in BH^p} \left(\int_{\mathbb{D}} |B_n f|^q dS \right)^{1/q}. \tag{7}$$

Applying the inequality between geometric and arithmetic means to the integral

$$T_r(f) = \int_{T_r} |B_n f(z)|^q |dz|$$

and putting $f(z) \equiv 1$ we obtain the estimate from below. The estimate from above is a consequence of (7). One should put $B_n(z) = z^n$, in (7), apply the Hölder inequality and well-known fact that $T_r(f)$ increase monotonically as $r \uparrow 1^-$.

We now return to proof of Theorem 2. Notice that the behavior of measure μ has no influence on asymptotics of n-widths. Indeed Gelfand n-widths are additive:

$$d^{n+m-1}(J_1 + J_2) \leq d^n(J_1) + d^m(J_2) \tag{8}$$

(see [4] p. 171). Further for measures μ with $\operatorname{supp}\mu \subset \mathbb{D}_q = \{z \mid |z| < q\}$, $q < 1$ there is an obvious estimate

$$d^n(J : H^p \to \mathcal{L}_q(\mu)) = O(q^n).$$

(See [5] for details).

Let ϕ be the weight of the form

$$\phi(z) = \begin{cases} |F'(z)|^{2-q/p}, & z \in \mathbb{D} \setminus \mathbb{D}_{r_0} \\ 0, & z \in \mathbb{D}_{r_0} \end{cases} \tag{9}$$

$r_0 < 1$ where F is a conformal map of $\mathbb{D} \setminus g_1$ onto $\mathbb{D} \setminus g_2$, g_1, g_2 being simply connected domains in $\mathbb{D}$, $g_2 \subset \mathbb{D}_{r_0}$, $F(\mathbb{T}) = \mathbb{T}$. We now observe that Theorem 2 reduces for weights (9) to the following formula

$$d_n(J) \sim (nq)^{-1/q}, \qquad n \to \infty.$$

Let us consider the diagram:

$$
\begin{array}{ccc}
H^p(\mathbb{D}\backslash g_1) & \xrightarrow{\;\;J\;\;} & \mathcal{L}_q(\mathbb{D},\phi dS) \\[4pt]
\Big\downarrow{\phi} & & \Big\uparrow{\tau} \\[4pt]
H^p(\mathbb{D}\backslash g_2) & \xrightarrow{\;\;\tilde{J}\;\;} & \mathcal{L}_q(\mathbb{D},\tilde{\phi}dS))
\end{array}
$$

where $\tilde{J}$ denotes the embedding operator and π,τ are defined as follows

$$(\pi f)(z) = f(\Phi(z))\cdot(\Phi'(z))^{1/p},$$
$$(\tau g)(z) = g(F(z))\cdot(F'(z))^{1/p},$$

Φ being the invers transform for F. The weight $\tilde{\phi}$ equals 1 near $\mathbb{T}$ and vanishes near the boundary of g_2. The existence of the branches of $(F')^{1/p},(\Phi')^{1/p}$ can be proved by Erochins Theorem [6] on conformal maps. This theorem says that any conformal map between double connected domains can be factored as a product of two conformal maps of simply connected domains. For the corresponding functions $(F')^{1/p},(\Phi')^{1/p}$ the previous diagram is commutative. Therefore we have $d_n(J) = d_n(\tilde{J})$. Instead of operator $\tilde{J}$ we can consider the operator J_0 because $\tilde{\phi}(z) \equiv 0$ in $\mathbb{D}$. So

$$d_n(\tilde{J}) \sim d_n(J_0) = (nq+2)^{-1/q}$$

and Theorem 2 is proved for weights of the form (9). If $\phi(z)$ is positive and continuous near $\mathbb{T}$ then we consider the weight $\bar{\phi}(z)$ such that $\phi(z) = C\bar{\phi}(z)$ and

$$\frac{1}{2\pi}\int_0^{2\pi}\left(\bar{\phi}\left(e^{it}\right)\right)^{\frac{p}{2p-q}}dt = 1. \qquad (10)$$

It is clear

$$C = \left(\frac{1}{2\pi}\int_0^{2\pi}\left(\phi\left(e^{it}\right)\right)^{\frac{p}{2p-q}}dt\right)^{\frac{2p-q}{p}}.$$

In order to finish the proof we need the following lemma on approximation.

LEMMA 2. Let $\bar{\phi}$ be a positive continuous weight satisfying (10). Then for any positive $\epsilon > 0$ there exist a neighbourhood E of $\mathbb{T}$ and a weight ϕ_ϵ of the form (9) such that

$$1 - \epsilon \leq \bar{\phi}/\phi_\epsilon \leq 1 + \epsilon, \qquad z \in E.$$

See [5] for proof of Lemma2.

References

[1] S.Fisher and C.Micchelli, *The n-widths of sets of analytic functions*, Duke Math.J. **47** no. 4 (1980), pp. 789–801.

[2] J.E.Andersson, *Optimal quadrature H^p-functions*, Matem.Zametki **172** no. 1 (1980), pp. 55–62. (In Russian)

[3] O.G.Parfenov, *Gelfand n-widths of the unit ball of H^p in weighted spaces*, Matem.Zametki **37** no. 2 (1985), pp. 171–175. (In Russian)

[4] A.Pietsch, *Operatorenidealen*, Berlin, VEB, 1978.

[5] O.G.Parfenov, *The estimates of singular numbers of the Carleson embedding operator*, Matem.Sbornik **131(173)** no. 4(2) (1986), pp. 501–518. (In Russian)

[6] V.D.Erochin, *Best linear approximation of functions having analytic continuation from given continuum onto given domain*, Uspechi Math. Nauk **23** no. 1(139) (1968), pp. 91–133. (In Russian)

St.Petersburg University
St.Petersburg Russia
St.Petergof, Biblioteschnja pl.2

ON THE CONVERGENCE
OF
BIEBERBACH POLYNOMIALS IN DOMAINS WITH INTERIOR ZERO ANGLES

I.E. Pritsker

1. Introduction

Let G be a domain in the complex plane $\mathbb{C}$ with Jordan boundary L, $0 \in G$. The Riemann function $\phi : G \to D$ maps G conformally onto the disk $D = \{w : |w| < r_0\}$ and is normalized by $\phi(0) = 0, \phi'(0) = 1$.

A polynomial $\pi_n(z)$ is called a Bieberbach polynomial if it minimizes the integral

$$\iint_G |P_n'(z)|^2 \, d\sigma_z$$

in the class of all polynomials $P_n(z)$ of degree $\leq n$ with normalization $P_n(0) = 0, \; P_n'(0) = 1$. It is well known that $\pi_n(z)$ also minimizes the integral

$$\iint_G |\phi'(z) - P_n'(z)|^2 \, d\sigma_z$$

in the same class. The last property implies the uniform convergence of Bieberbach polynomials to $\phi(z)$ on compact subsets of G.

The first result on the convergence of Bieberbach polynomials to the Riemann function in the closure of domain G was obtained by M.V.Keldysh under additional assumptions on the smoothness of L [1]. He has also constructed an example of domain, the boundary of which is analytic with exception of a single point and the sequence of Bieberbach polynomials diverges at this point. Later many authors obtained substantially weaker conditions on the boundary of G which guarantee the uniform convergence of the Bieberbach polynomials on G. The most known results of such type are due to V.V.Andrievskii who proved the convergence of Bieberbach polynomials in domains with quasiconformal boundaries [2] as well as with piecewise quasiconformal boundaries with certain zero angles [3,4]. D.Gaier proved variuos estimates of the rate of uniform approximation by Bieberbach polynomials in domains with quasiconformal boundaries [5].

2. The main results

Here we consider domains with interior zero angles. Let L be piecewise smooth and $\lambda\pi(0 < \lambda \leq 2)$ be the smallest exterior angle at the joint point of smooth arcs. We consider the wedge $K = \{z = x + iy : -f(x) < y < f(x), 0 < x < c\}$, where $f(x)$ is a positive, continuous and monotonically increasing function on $[0,c]$. We are especially interested in functions f, satisfying $f(x) = o(x)$ when $x \to 0+$. Let us assume that we can place the wedge K with vertex at every joint point of L so that $K \subset G$. Function $f(x)$ defines the largest possible order of touching at the interior zero angles of G.

If g is continuous on G, then

$$\|g\|_{C(\overline{G})} = \max_{z \in G} | g(z) |$$

THEOREM 1. *Suppose that for every $\epsilon > 0$, $x^{1+\epsilon} = o(f(x))$, as $x \to 0+$. Then for any $s < \min(\frac{1}{2}, \frac{\lambda}{2-\lambda})$.*

$$\|\phi - \pi_n\|_{C(\overline{G})} \leq \frac{C_1}{n^s} \tag{1}$$

Remark 1. Estimate (1) extends the result of [5, Th. 1] to domains with the described interior zero angles.

Remark 2. It was proved in [3, Th. 3] that if $f(x) = x(-\ln x)^\beta$, $(\beta > 0)$, then

$$\|\phi - \pi_n\|_{C(\overline{G})} \leq \frac{C_2}{n^{\epsilon_1}}$$

where $\epsilon_1 > 0$ is a certain number. However, this result doesn't contain the estimates of value ϵ_1.

THEOREM 2. *If $f(x) = C_3 x^\alpha$ $(C_3 > 0, \alpha > 1)$, then for any $s < 1 - \alpha + \min(\frac{2-\alpha}{2\alpha}, \frac{\lambda}{2-\lambda})$*

$$\|\phi - \pi_n\|_{C(\overline{G})} \leq \frac{C_4}{n^s} \tag{2}$$

THEOREM 3. *If $f(x) = C_3 x^\alpha$ $(C_3 > 0, \alpha > 1)$ and L has only interior zero angles at the joint points, then for any $s < \frac{3}{2} - \alpha$*

$$\|\phi - \pi_n\|_{C(\overline{G})} \leq \frac{C_5}{n^s} \tag{3}$$

Remark 3. It is easy to see that $\alpha < \frac{3}{2}$ is permissible in th 3 for the convergence of Bieberbach polynomials. This result is better then those of [4, Th. 4] and [6].

3. The method and auxiliary results

To prove Theorems 1–3 we will use the method of V.V.Andrievskii [3] and some estimates of D.Gaier [5].

With the notation

$$\|g\|_{L_2^1(G)} = \left[\iint_G |g'(z)|^2 \, d\sigma_z \right]^{\frac{1}{2}}$$

we have the following lemmas.

LEMMA 1. *Suppose* $x^{1+\epsilon} = o(f(x))$ *as* $x \to 0+$ *for any* $\epsilon > 0$. *Then for any polynomial* $P_n(z)$, $P_n(0) = 0$,

$$\|P_n\|_{C(\overline{G})} \le C_6 n^\epsilon \|P_n\|_{L_2^1(G)}$$

LEMMA 2. *If* G *satisfies the conditions of Th. 2, then for any polynomial* $P_n(z), P_n(0) = 0$,

$$\|P_n\|_{C(\overline{G})} \le C_7 n^{\alpha-1} \|P_n\|_{L_2^1(G)}$$

Lemmas 1–2 follows from [7, Th. 2].

LEMMA 3. *Under the assumptions of Th. 1 for any* $p < \min(\frac{1}{2}, \frac{\lambda}{2-\lambda})$ *we have*

$$\|\phi - \pi_n\|_{L_2^1(G)} \le \frac{C_8}{n^p}$$

LEMMA 4. *If* G *satisfies the conditions of Th. 2, then for any* $p < \min(\frac{2-\alpha}{2\alpha}, \frac{\lambda}{2-\lambda})$

$$\|\phi - \pi_n\|_{L_2^1(G)} \le \frac{C_9}{n^p}$$

LEMMA 5. *If* G *satisfies the conditions of Th. 3, then for any* $p \le \frac{1}{2}$

$$\|\phi - \pi_n\|_{L_2^1(G)} \le \frac{C_{10}}{n^p}$$

Combining Lemmas 1–5 with [3, Lemma 15], one can easily obtain Theorems 1–5.

Using the same method we can get the similar results on the convergence of Bieberbach polynomials in domains with piecewise quasiconformal boundaries and interior zero angles at the joint points of quasiconformal arcs.

References

[1] Keldysh M., *Sur l'approximation en moyenne quadratique des fonctions analytiques*, Mat. Sb. **5** no. 2 (1939), pp. 391–401.

[2] Andrievskii V.V., *Convergence of Bieberbach polynomials in domains with quasiconformal boundary*, Ukrainian Math. J., **35** no. 3 (1983), pp. 273–277. (In Russian)

[3] Andrievskii V.V., *Uniform convergence of Bieberbach polynomials in domains with piecewise-quasiconformal boundary*, In: Theory of Mappings and Approximation of Functions, Naukova Dumka, Kiev, 1983, pp. 3–18. (In Russian)

[4] Andrievskii V.V., *Uniform convergence of Bieberbach polynomials in domains with zero angles*, Dokl. Akad. Nauk Ukrain. SSR Ser. **A** no. 4 (1982), pp. 3–5. (In Russian)

[5] Gaier D., *On the convergence of the Bieberbach polynomials in regions with corners*, Constr. Approx. **4** (1988), pp. 289–305.

[6] Kulikov I.V., *Computational method of approximation by polynomials in Jordan domains satisfying the segment condition Rostov-na-Donu*, VINITI **2853** (1988), p. 251. (In Russian)

[7] Pritsker I.E., *The order relationships between the polynomial norms in complex domains*, Ukrainian Math. J., (to appear). (In Russian)

Institute of Applied Mathematics and Mechanics
Ukrainian Academy of Sciences
Donetsk USSR

DUALITY PRINCIPLE IN LINEARIZED
RATIONAL APPROXIMATION

Boris Shekhtman

We prove a duality theorem for linearized rational approximation similar to the theorem for rational approximation proved by G. Gierz and the author earlier.

1. Introduction

In this article we prove a duality theorem for the linearized rational approximation. This result is in the same spirit as the duality theorem for rational approximation (cf. [2] and [3]).

Let $C(K)$ be the space of real-valued continuous functions on a compact Hausdorff space K. Let G and H be subspaces of $C(K)$. We form the set $R(H,G) = \{g/h : g \in G, h \in H, h(k) > 0, \forall k \in K\}$. Using the notation $H^\perp = \{\mu \in [C(K)]^* : \mu(h) = 0 \text{ for all } h \in H\}$ we can prove

THEOREM 1.1. (cf. [2] and [3]) *The following are equivalent:*

    (i) $\text{dist}(f, R(H,G)) > d$

    (ii) *There exists $\mu \in G^\perp$, $\nu \in H^\perp$ such that $f\mu + \nu \geq d|\mu|$ where $|\mu|$ is the absolute variation of the measure μ.*

In this paper, we deal with another set: $LR(H,G)$. We define the number $\text{dist}(f, LR(H,G))$ to be the least number $\varepsilon > 0$ such that $\|fh - g\| \leq \varepsilon$ for some $h \in H$; $g \in G$; $\|h\| > 1$.

In the next section we prove the main duality theorem for this "linearized rational approximation." One of the advantages of dealing with the linearized version is that the theorem holds true not only for real-valued continuous functions but also for complex-valued L_p spaces.

In the interest of generality we prove this theorem for operators A acting between two Banach spaces X and Y. Hence we will be interested in the conditions on the subspaces $H \subset X$ and $G \subset Y$ that guarantee the existence of $x \in H$, $y \in G$ with $\|Ax - y\| < \varepsilon$; $\|x\| > 1$.

While the duality theorem can be proved in this level of generality, our primary interest results from the case when $\mathbf{X} \doteq \mathbf{Y}$ is a function space and $A = A_f$ is a multiplication operator: $A_f x = f \cdot x$.

In the third section, we provide some applications of the main theorem as well as some comparisons between rational and linearized rational approximation.

2. The Main Theorem

Let H, G be linear subspaces of Banach spaces $\mathbf{X}$ and $\mathbf{Y}$. Let A be a linear operator from $\mathbf{X}$ into $\mathbf{Y}$.

We say $A \in LR(H, G)$ if for every $\varepsilon > 0$ there exist $h \in H$, $g \in G$ such that

$$\|h\| > 1 \quad \text{and} \quad \|Ah - g\| < \varepsilon.$$

We further define $\text{dist}(f, LR(H, G)) := \inf\{\varepsilon : \|Ah - g\| < \varepsilon \text{ and } h \in H, g \in G; \|h\| > 1\}$.

THEOREM 2.1. *Let A, H, G be as above. Then the following are equivalent*

(i) $\text{dist}(A, LR(H, G)) \geq d$
(ii) *For every $\sigma \in \mathbf{X}^*$ there exist $\mu \in G^{\perp}$; $\nu \in H^{\perp}$ such that*
$$A^* \mu + \nu = \sigma \quad \text{and} \quad \|\sigma\| \geq d\|\mu\|.$$

Proof. Assume (i) and pick $\varepsilon < d$. For an arbitrary $\sigma \in \mathbf{X}^*$ consider the subset of $\mathbf{X} \times \mathbf{Y}$ defined by $W(\sigma, \varepsilon) = \{(x, y) \in \mathbf{X} \times \mathbf{Y} : \|Ax - y\| < \varepsilon, \ \sigma(x) > 1\}$. Clearly $W(\sigma, \varepsilon)$ is a convex cone with a non-empty interior. Our assumption implies $W(\sigma, \varepsilon) \cap (H \times G) = \emptyset$.

Hence by the Hahn-Banach theorem, there exists a non-trivial functional $\varphi \in (\mathbf{X} \times \mathbf{Y})^*$ such that

$$\varphi(w) \geq 0 \quad \text{for all} \quad w \in W$$

and

$$\varphi \in (H \times G)^{\perp}.$$

Since every $\varphi \in (\mathbf{X} \times \mathbf{Y})^*$ can be represented as a pair of functionals $(\nu, \mu) \in \mathbf{X}^* \times \mathbf{Y}^*$ with $(\nu, \mu)(x, y) = \nu(x) + \mu(y)$ the conditions (2.5), (2.6) are translated into the following.

There exist functionals $\nu \in H^{\perp}$, $\mu \in G^{\perp}$ such that

$$\|\mu\| + \|\nu\| > 0$$

$$\nu(x) + \mu(y) \geq 0 \quad \text{for all} \quad (x, y) \in W(\sigma, \varepsilon).$$

Next we pick $x_0 \in \mathbf{X}$ with $\sigma(x_0) > 1$. Then for every $z \in \ker \sigma$ and every natural number n, and for every $y_0 \in Y$ with $\|y_0\| \leq 1$ we have $(x_0 + nz, Ax_0 + nAz - \varepsilon y_0) \in W(\sigma, \varepsilon)$.

Hence by (2.8) $\nu(x_0) + \mu(Ax_0) + n\nu(z) + n\mu(Az) - \varepsilon\mu(y_0) \geq 0$ or

$$(\nu + A^*\mu)(x_0) + n(\nu + A^*\mu)(z) \geq \varepsilon\mu(y_0).$$

We now claim that $(\nu + A^*\mu)(z) = 0$ for all $z \in \ker\sigma$. Indeed if it was not so we could choose either z or $-z$ and n large enough so that (2.9) fails.

Hence $\ker(\nu + A^*\mu) \supset \ker\sigma$. It is now easy to see from (2.9) that $\nu + A^*\mu \neq 0$ if $\mu \neq 0$ and from (2.7) we have $\nu + A^*\mu \neq 0$ if $\mu = 0$. Thus $\operatorname{codim}\ker(\nu + A^*\mu) = 1 = \operatorname{codim}\ker\sigma$ which implies $\ker(\nu + A^*\mu) = \ker\sigma$ and

$$\nu + A^*\mu = \alpha\sigma$$

for some non-zero constant α.

From (2.9) we have $\varepsilon\mu(y_0) \leq \alpha\sigma(x_0)$ for all y_0 such that $\|y_0\| < 1$ and for all x_0 such that $\sigma(x_0) > 1$. Maximizing this inequality over all such x_0 and minimizing over all such y_0 we have $\varepsilon \left\|\left(\frac{\mu}{\alpha}\right)\right\| \leq \|\sigma\|$ and from (2.10) $\left(\frac{\nu}{\alpha}\right) + A^*\left(\frac{\mu}{\alpha}\right) = \sigma$ which is precisely (ii).

We now assume (ii) and prove that $\operatorname{dist}(A, LR(H,G)) \geq d$. Indeed let $\|Ah - g\| < d$; $(h,g) \in H \times G$ and $\|h\| > 1$. Then there exists $\sigma \in \mathbf{X}^*$ such that $\|\sigma\| = 1$ and $\sigma(h) > 1$. By (ii) there exists $\mu \in G^\perp$; $\nu \in H^\perp$ such that $A^*\mu + \nu = \sigma$ and $\|\sigma\| \geq d\|\mu\|$. Then $\|\sigma\| = 1 < (A^*\mu + \nu)(h) = \mu(Ah) = \mu(Ah - g) \leq \|\mu\|d$ which contradicts (2.4).

COROLLARY 1. Let $A \notin LR(H,G)$. Then$A^*G^\perp + H^\perp = \mathbf{X}^*$.

I do not know the meaning of the next corollary but it sounds like a peculiar form of a fixed point theorem.

COROLLARY 2. Let G, H be subspaces of $C(K)$. Let $f \in C(K)$ and suppose that for every regular Borel measure σ on K there exist measures $\mu \perp G$, $\nu \perp H$ such that $f\mu + \nu = \sigma$; $\|\sigma\| > \varepsilon\|\mu\|$. Then there exist measures $\mu \in G^\perp$, $\nu \in H^\perp$ such that $f\mu + \nu \geq \varepsilon|\mu|$.

Proof. The proof follows from Theorem 1.1 and from $d(f, LR(H,G)) \leq d(f, R(H,G))$.

3. Some Examples

Now we turn our attention to the linearized rational approximation. We thus assume that $G = H = E$ is a subspace of a real or complex $C(K)$ space and the operator A is given by a function $f \in C(K)$: $A_f x = f \cdot x$. We use the symbols $R(E)$ and $LR(E)$ to denote $R(E,E)$ and $LR(E,E)$ as subsets of $C(K)$.

Numerous examples given in [2], [3], and [4] show how one can construct a subspace $E \subset C(K)$ of codimension 1 so that $R(E)$ is *not dense* in $C(K)$.

PROPOSITION 3.1. *Let H and G be subspaces of $\mathbf{X}$ and $\mathbf{Y}$ of finite codimension. Let $\dim \mathbf{X} = \infty$. Then $LR(H,G) = \mathcal{L}(\mathbf{X}, \mathbf{Y})$ – the space of all linear operators from $\mathbf{X}$ into $\mathbf{Y}$.*

Proof. Let $A \notin LR(H,G)$. Then by Theorem 2.1 $\mathbf{X}^* = H^\perp + A^* G^\perp$ which is impossible since the dimensions of $H^\perp$ and $G^\perp$ are finite.

Turning to another extreme, we consider the case when $E \subset C(K)$ is a finite-dimensional space. It is easy to see that $R(E)$ is dense in $C(K)$ implies K is finite. For linearized rational approximation, the corresponding result is:

PROPOSITION 2. *K has an isolated point if and only if there exists a finite-dimensional subspace $E \subset C(K)$ such that $LR(E) = C(K)$.*

Proof. Let $k_0 \in K$ be an isolated point and let $E \subset C(K)$ be the span of the function ℓ. $\ell(k) = \begin{cases} 2 & \text{if} \quad k = k_0 \\ 0 & \text{if} \quad k \neq k_0. \end{cases}$ Then for any $f \in C(K)$ we have $\|f \cdot \ell - f(k_0) \cdot \ell\| = 0$. Conversely, let E be any finite-dimensional subspace of $C(K)$ and suppose K has no isolated points. Then every open set $\mathcal{O} \subset K$ contains infinitely many points and hence there exists a function f such that

$$f|\mathcal{O} \notin R(E|\mathcal{O}) \quad \text{for all open} \quad \mathcal{O} \subset K.$$

On the other hand, if $\ell_n, \ell'_n \in E$ are such that $\|f\ell_n - \ell'_n\| < \frac{1}{n}$; $\|\ell_n\| > 1$ we have

$$f \cdot \frac{\ell_n}{\|\ell_n\|} - \frac{\ell'_n}{\|\ell_n\|} \to 0.$$

Since E is finite-dimensional and $\|\ell_n / \|\ell_n\|\| < 1$ we can assume, by passing to a subsequence, that $\frac{\ell_n}{\|\ell_n\|} \to \ell$. By (3.2), it is implied that $\ell'_n/\|\ell'_n\| \to \ell' \in E$. Hence $f\ell = \ell'$. Since $\|\ell\| \geq 1$ there exists an open set $\mathcal{O}$ such that $\ell|\mathcal{O} > 0$ and hence $f|\mathcal{O} \in R(E|\mathcal{O})$ which contradicts the assumption (3.1).

Next we give examples of subspaces E of infinite dimension and codimension such that $LR(E) = C(K)$ and such that $LR(E) \neq C(K)$.

We will deal with the space $C_{[-\pi,\pi]}$ of 2π-periodic continuous complex-valued functions. Let $\Lambda \subset Z$ be a Sidon set (for instance, $\Lambda = \{4^n : n \in N\}$).

PROPOSITION 3.3. *Let $E = \mathrm{span}\{e^{i\lambda\theta}, \lambda \in \Lambda\}$. Then $LR(E) \neq C_{[-\pi,\pi]}$.*

Proof. For any Sidon set Λ there exist positive constants c_1, c_2 such that

$$c_2 \sum_{\lambda \in \Lambda} |a_\lambda| \leq \left\| \sum_{\lambda \in \Lambda} a_\lambda e^{i\lambda\theta} \right\| \leq c_1 \sum_{\lambda \in \Lambda} |a_\lambda|.$$

Pick $\lambda_0 \notin \Lambda$ such that $\Lambda \cap (\Lambda + \lambda_0) = \emptyset$. Since the union of two Sidon sets is a Sidon set (cf. [1]) the set $\Lambda' = \Lambda \cup (\Lambda + \lambda_0)$ is Sidon. Thus if

$$\left\| e^{i\lambda_0} \left(\sum_{\lambda \in \Lambda'} a_\lambda e^{i\lambda\theta} \right) - \sum_{\lambda \in \Lambda} b_\lambda e^{i\lambda\theta} \right\| < \varepsilon$$

then $\sum |a_\lambda| < c_2\varepsilon$. On the other hand, if $\left\| \sum_{\lambda \in \Lambda} a_\lambda e^{i\lambda\theta} \right\| > 1$ then $\sum |a_\lambda| > c_1$. Now we can choose ε so small that these two conditions are inconsistent.

On the other hand

PROPOSITION 3.4. *Let Λ be a Sidon set and let $E = \mathrm{span}\left\{ e^{in\theta},\ n \notin \Lambda \right\}$.*

Then $LR(E) = C_{[-\pi,\pi]}$.

Proof. It folows from a theorem of Bochner (cf. [1]) that $E^\perp \subset L_1(m)$ where m is Lebesgue measure. In particular, $E^\perp$ is separable. Hence $fE^\perp + E^\perp \neq (C_{[-\pi,\pi]})^*$ since the space on the left is separable and the one on the right is not.

Remark 3.5 This proposition is in remarkable contrast with the similar situation for $R(E)$. Let $\Lambda = \{0,3,4,5,\ldots\}$. Let $E = \mathrm{span}\{\cos\lambda\theta : \lambda \in \Lambda\} \subset C_{[0,\pi]}$. Then $R(E)$ is not dense in $C_{[0,\pi]}$.

Indeed if $f = 2\cos\theta$; $\mu = \cos\theta$; $\nu = -\cos 2\theta$ we have $f\mu + \nu = 1 \geq |\mu|$. Hence by Theorem 1.1, we have the desired conclusion.

Problem 3.6 What are the sets $\Lambda \subset \mathbf{N}$ such that $LR(E) = C_{[-\pi,\pi]}$ where $E = \mathrm{span}\{e^{i\lambda\theta},\ \lambda \in \Lambda\}$?

More exotic examples of such sets Λ can be easily constructed using results in [5] where several translation-invariant subspaces $E \subset C_{[-\pi,\pi]}$ with $E^\perp$ being separable are given.

It is probably true that for every $\Lambda \subset \mathbf{N}$ either $LR(\mathrm{span}\{e^{in\theta} : n \in \Lambda\})$ or $LR(\mathrm{span}\{e^{in\theta} : n \notin \Lambda\})$ is dense in $C_{[-\pi,\pi]}$.

Problem 3.7 Let $E \subset C(K)$ such that for every $\mu \in E^\perp$ $\mathrm{supp}\,(\mu^+) \cap \mathrm{supp}\,(\mu^-) \neq \emptyset$. Does it follow that $LR(E) = C(K)$?

References

[1] R.E. Edwards, *Fourier Series, A Modern Introduction*, vol. 2, Springer-Verlag, 1982.

[2] G. Gierz and B. Shekhtman, *On Duality in Rational Approximation*, Rocky Mountain Journal **19** (1988), pp. 137–143.

[3] G. Gierz and B. Shekhtman, *Duality Principle for Rational Approximation*, Pacific Journal of Math **125** (1986), pp. 79–90.

[4] G. Gierz and B. Shekhtman, *On Approximation by Rationals from a Hyperplane*, Proc. AMS **96** (1986), pp. 452–454.

[5] S. Kwapien and A. Pelczynski, *Absolutely Summing Operators and Translation Invariant Subspaces on Compact Abelian Groups*, Math. Nachr. **94** (1980), pp. 303–340.

[6] D.J. Newman, *Approximation with Rational Functions*, CBMS, Regional Conf. Ser. no. 41 (1979), Providence R.I.

Dept. of Mathematics
USF, Tampa 33620
USA

UNIVERSALITY
OF THE FIBONACCI CUBATURE FORMULAS

V.N. Temlyakov

Introduction

Let $\{b_n\}_{n=0}^{\infty}$, $b_0 = b_1, b_n = b_{n-1} + b_{n-2}, n \geq 2$, be the Fibonacci numbers. For continuous functions of two variables which are 2π-periodic in each variable we define cubature formulas

$$\Phi_n(f) = b_n^{-1} \sum_{\mu=1}^{b_n} f(2\pi\mu/b_n, 2\pi\{\mu b_{n-1}/b_n\}),$$

which will be called Fibonacci cubature formulas. Here $\{a\}$ denotes the fractional part of the number a. The Fibonacci cubature formulas Φ_n are the special case of parallelepipedal number-theoretic cubature formulas introduced by N.M. Korobov [1]. The first results on cubature formulas Φ_n can be found in N.S. Bakhvalov [2], Hua Loo Keng, Wang Yuan [3] and N.M. Korobov [4]. The further detailed investigation of the Fibonacci cubature formulas was carried out in the author's papers [5–7] (see also the book [8]).

We now state the known results. We consider the class $\mathbf{MW}_{p,\alpha}^{\mathbf{r}}$, $\mathbf{r} = (r_1, r_2)$, $l \leq p \leq \infty$, $\alpha = (\alpha_1, \alpha_2)$ of functions $f(\mathbf{x}), \mathbf{x} = (x_1, x_2)$, which have the following integral representation

$$f(\mathbf{x}) = (2\pi)^{-2} \int_{\mathbb{T}^2} \varphi(\mathbf{x} - \mathbf{t})F_{\mathbf{r}}(\mathbf{t}, \alpha)dt, \| \varphi \|_p \leq 1,$$

where

$$F_{\mathbf{r}}(\mathbf{t}, \alpha) = F_{r_1}(t_1, \alpha_1)F_{r_2}(t_2, \alpha_2),$$

$$F_\rho(y, \beta) = 1 + 2\sum_{k=1}^{\infty} k^{-\rho} \cos(ky - \beta\pi/2).$$

We denote $r = \min(r_1, r_2)$. It is known (see [9], Ch.2) that the restriction $r > 1/p$ is necessary and sufficient for $\mathbf{MW}^r{}_{p,\alpha}$ be embedded into the space of continuous functions. We first state the results for $r_1 = r_2$ (see [5–7]). Given a functional class F we denote

$$\Phi_n(F) = \sup_{f\in F} |\Phi_n(f) - (2\pi)^{-2} \int_{\mathbb{T}^2} f(\mathbf{x})dx|.$$

THEOREM 1. *Let* $1 \leq p \leq \infty$, $\mathbf{r} = (r, r)$, $r > 1/p$. *Then for any* α *we have*

$$\Phi_n(\mathbf{MW^r}_{p,\alpha}) \asymp \begin{cases} b_n^{-r} \log b_n, & p = 1, r > 1 \\ b_n^{-r}(\log b_n)^{1/2}, & 1 < p \leq \infty, r > \max(1/p, 1/2) \\ b_n^{-r}(\log b_n)^{1-r}, & 2 < p \leq \infty, 1/p < r < 1/2 \\ b_n^{-r}((\log b_n)(\log \log b_n))^{1/2}, & 2 < p \leq \infty, r = 1/2. \end{cases}$$

Given a functional class F let

$$\Lambda_m(f, \xi) = \sum_{j=1}^m \lambda_j f(\xi^j),$$

$$\Lambda_m(F, \xi) = \sup_{f \in F} |\Lambda_m(f, \xi) - (2\pi)^{-2} \int_{\mathbb{T}^2} f(\mathbf{x}) d\mathbf{x}|, \tag{1}$$

$$\delta_m(F) = \inf_{\substack{\lambda_1, \dots, \lambda_m \\ \xi_1, \dots, \xi_m}} \Lambda_m(F, \xi).$$

The number $\delta_m(F)$ is called the optimal error of cubature formulas with m points.

As a consequence of Theorem 1 and of Theorem 1.1 of [10] we have

$$\delta_m(\mathbf{MW^r}_{p,\alpha}) \gg m^{-r}(\log m)^{1/2}, \quad 1 < p < \infty, r > 1/p. \tag{2}$$

The second relation in Theorem 1 and (2) show that the Fibonacci cubature formulas Φ_n provide the optimal error in the sense of order inside the class of cubature formulas with b_n points if $1 < p < \infty$, $r > \max(1/p, 1/2)$. In the first section we prove that for $r_1 \neq r_2$ the Fibonacci cubature formulas provide the optimal error in the sense of order for all $1 \leq p \leq \infty$ and $r > 1/p$.

THEOREM 2. *Let* $1 \leq p \leq \infty$, $\mathbf{r} = (r_1, r_2)$, $r_1 \neq r_2$, $r = \min(r_1, r_2)$, $r > 1/p$. *Then for any* α *we have*

$$\Phi_n(\mathbf{MW^r}_{p,\alpha}) \asymp \delta_{b_n}(\mathbf{MW^r}_{p,\alpha}) \asymp b_n^{-r}.$$

In the second section we investigate the error of the Fibonacci cubature formulas for the Sobolev classes $\mathbf{SW^R}_{p,\alpha}$ and the Nikol'skii classes $\mathbf{NH^R_p}$ for various vectors $\mathbf{R}$ (see the definition of these classes in § 2). These classes differ from the classes $\mathbf{MW^r}_{p,\alpha}$ fundamentally. But it turned out that the Fibonacci cubature formulas yield the optimal error in the sense of order for all such classes too.

THEOREM 3. *Let* F_p^R *denote one of the classes* $SW_{p,\alpha}^R$ *or* NH_p^R, $1 \leq p \leq \infty$, $R = (R_1, R_2)$, $g(R) = (R_1^{-1} + R_2^{-1})^{-1} > 1/p$. *Then*

$$\Phi_n(F_p^R) \asymp \delta_{b_n}(F_p^R) \asymp b_n^{-g(R)}.$$

Theorems 2 and 3 and Theorem 1 with (2) show that the Fibonacci cubature formulas provide the optimal error in the sense of order for a wide set of various functional classes. This property demonstrates a universal character of the Fibonacci cubature formulas.

1. Proof of Theorem 2

It is easy to see that the lower estimate $\delta_m(MW_{p,\alpha}^r) \gg m^{-r}$ follows from the lower estimate in the one-dimensional case

$$\delta_m(W_{p,\beta}^r) \gg m^{-r}. \tag{1.1}$$

In Section 2 we prove the lower estimate for the Sobolev classes $SW_{p,\alpha}^R$, which implies (1.1).

We now prove the upper estimate $\Phi_n(MW_{p,\alpha}^r) \ll b_n^{-r}$. We prove it for a wider class MH_p^r consisting of functions with the following restriction on mixed differences:

$$MH_p^r = \{f(x) : \| \Delta_{t_1}^l \Delta_{t_2}^l f(x) \|_p \leq |t_1|^{r_1} |t_2|^{r_2}, l = \max([r_1], [r_2]) + 1\}.$$

Let $V_m(t)$ be the de la Vallée-Poussin kernel of order $2m - 1$ and

$$A_0(t) \equiv 1; \quad A_1(t) = 2V_1(t) - 1,$$
$$A_s(t) = 2(V_{2^s-1}(t) - V_{2^s-2}(t)), \quad s \geq 2;$$
$$A_s(x) = A_{s_1}(x_1)A_{s_2}(x_2), \quad s = (s_1, s_2);$$

$$A_s(f) = f * A_s.$$

It is known (see [9] p.33) that for $f \in MH_p^r$ we have

$$\| A_s(f) \|_p \ll 2^{-(r,s)}. \tag{1.2}$$

Let $L(n) = \{k = (k_1, k_2) : k_1 + k_2 b_{n-1} \equiv 0 (\mathrm{mod}\, b_n)\}$. We consider the following functions

$$A_s(L(n), x) = \sum_{k \in L(n)} \hat{A}_s(k) e^{i(k,x)}.$$

We formulate a consequence of the results of [6] (see p. 332).

LEMMA 1.1. *There is an absolute constant $\beta > 0$ such that for all $s = (s_1, s_2) : 0 < s_1 + s_2 < m = [\log \beta b_n]$*

$$A_s(L(n), \mathbf{x}) \equiv 0$$

and for $s : \| s \|_1 = s_1 + s_2 \geq m$ we have

$$\| A_s(L(n), \mathbf{x}) \|_q \ll 2^{(\|s\|_1 - m)(1 - 1/q)}. \tag{1.3}$$

It is easy to see that without loss of generality we may assume that f is a trigonometric polynomial. Considering (1.2) and (1.3) we find

$$|\Phi_n(f) - \hat{f}(0)| = \left| \sum_{k \in L(n) \backslash 0} \hat{f}(\mathbf{k}) \right| = \left| \left(f(\mathbf{x}), \sum_{\|s\|_1 \geq m} A_s(L(n), \mathbf{x}) \right) \right| =$$

$$= \left| \sum_{\|s\|_1 \geq m} \sum_{u : \|u - s\|_\infty \leq 1} (A_u(f), A_s(L(n))) \right| \ll \sum_{\|s\|_1 \geq m} 2^{-(\mathbf{r}, s) + (\|s\|_1 - m)/p}.$$

Using Lemma C from [9] (p.9) we get $|\Phi_n(f) - \hat{f}(0)| \ll 2^{-rm} \asymp b_n^{-r}$. This completes the proof of the upper estimate.

We note that in fact we have proved the following more general statement.

THEOREM 1.1. *Let $\mathbf{MF}_p^{\mathbf{r}}$ denote one of the classes $\mathbf{MW}_{p,\alpha}^{\mathbf{r}}$ or $\mathbf{MH}_p^{\mathbf{r}}$. Then for $1 \leq p \leq \infty$, $\mathbf{r} = (r_1, r_2)$, $r_1 \neq r_2$, $r = \min(r_1, r_2)$, $r > 1/p$ we have*

$$\Phi_n(\mathbf{MF}_p^{\mathbf{r}}) \asymp \delta_{b_n}(\mathbf{MF}_p^{\mathbf{r}}) \asymp b_n^{-r}.$$

2. Proof of Theorem 3

We first give the definition of the classes $\mathbf{SW}_{p,\alpha}^{\mathbf{R}}$ and $\mathbf{NH}_p^{\mathbf{R}}$. Let $\mathbf{R} = (R_1, R_2)$, $1 \leq p \leq \infty$ and

$$\mathbf{SW}_{p,\alpha}^{\mathbf{R}} = \{f(\mathbf{x}) = (2\pi)^{-1} \int_0^{2\pi} \varphi_1(x_1 - t, x_2) F_{R_1}(t, \alpha_1) dt,$$

$$f(\mathbf{x}) = (2\pi)^{-1} \int_0^{2\pi} \varphi_2(x_1, x_2 - t) F_{R_2}(t, \alpha_2) dt, \| \varphi_j \|_p \leq 1, j = 1, 2\},$$

$$\mathbf{NH}_p^{\mathbf{R}} = \{f(\mathbf{x}) : \| \triangle_h^{l_j, j} f \|_p \leq |h|^{R_j}, l_j = [R_j] + 1, j = 1, 2\}$$

where $\triangle_h^{l, j}$ is the l-th difference with the step h in x_j.

It is known that the Sobolev classes $\mathbf{SW}_{p,\alpha}^{\mathbf{R}}$ are embedded into the Nikol'skii classes $\mathbf{NH}_p^{\mathbf{R}}$. Thus, it suffices to prove the upper estimate for

$\mathbf{NH}_p^{\mathbf{R}}$ and the lower estimate for $\mathbf{SW}_{p,\alpha}^{\mathbf{R}}$. We start with the proof of the upper estimate.

Clearly, without loss of generality we may assume that f is a trigonometric polynomial. Then using the notation of §1 we have

$$\Phi_n(f) = \sum_{k \in L(n)} \hat{f}(k). \qquad (2.1)$$

Let

$$A(\mathbf{R}, l, \mathbf{x}) = 4 \prod_{j=1}^{2} V_{[2^{\rho_j l}]}(x_j) - 4 \prod_{j=1}^{2} V_{[2^{\rho_j(l-1)}]}(x_j), l = 1, 2, \dots,$$

$$A(\mathbf{R}, 0, \mathbf{x}) = 4V_1(x_1)V_1(x_2)$$

where $\rho_j = g(\mathbf{R})/R_j, j = 1, 2$ and

$$A(\mathbf{R}, l, \mathbf{x})_{L_n} = \sum_{k \in L(n)} \hat{A}(\mathbf{R}, l, k) e^{i(k, \mathbf{x})}.$$

LEMMA 2.1. Let $l_0 = \max(l : A(\mathbf{R}, l) \in T(\beta b_n))$, where β is from Lemma 1.1 and $T(N)$ is the set of trigonometric polynomials with harmonics in

$$\Gamma(N) = \{k = (k_1, k_2) : \prod_{j=1}^{2} \max(|k_j|, 1) \le N\}.$$

Then $A(\mathbf{R}, 0)_{L(n)} = 1, A(\mathbf{R}, l)_{L(n)} = 0, l = 1, 2, \dots, l_0$, and for $l > l_0$

$$\| A(\mathbf{R}, l)_{L(n)} \|_q \ll (2^l / b_n)^{1-1/q}, 1 \le q \le \infty.$$

Proof. The first statement of the lemma follows directly from Lemma 1.1. Lemma 1.1 implies also that the number of nonzero terms of $A(\mathbf{R}, l)_{L(n)}$ is less (in order) than $2^l / b_n$ which gives the second statement for $q = \infty$.

Let us prove the estimate for $q = 1$. We make use of the function $G_N(\mathbf{x})$ defined in [6] (p. 332). Let l be fixed. We assume N to be so large that for all $k \in L(n)$ such that $\hat{A}(\mathbf{R}, l, k) \ne 0$ we have $\hat{G}(k) = 1$.
Then

$$A(\mathbf{R}, l)_{L(n)} = A(\mathbf{R}, l) * G_N \qquad (2.2)$$

It was proved in [6] that $\| G_N \|_1 \ll 1$. This relation and (2.2) imply

$$\| A(\mathbf{R}, l)_{L(n)} \|_1 \ll 1. \qquad (2.3)$$

which proves the inequality from Lemma 2.1 for $q = 1$.

The general case $1 \le q \le \infty$ follows from the previous cases by the inequality

$$\| f \|_q \le \| f \|_1^{1/q} \| f \|_\infty^{1-1/q}.$$

Remark. The number l_0 is such that $2^{l_0} \asymp b_n$. From (2.1) and Lemma 2.1 we obtain

$$\Phi_n(f) - \hat{f}(0) = \sum_{l > l_0} (f, A(\mathbf{R}, l)_{L(n)}). \tag{2.4}$$

We denote $A(f, \mathbf{R}, m) = f * A(\mathbf{R}, m)$. It is known (see for example [8], Ch.2, §3) that for $f \in NH_p^{\mathbf{R}}$ we have

$$\| A(f, \mathbf{R}, m) \|_p \ll 2^{-g(\mathbf{R})m}. \tag{2.5}$$

Lemma 2.1 and (2.5) yield

$$|(f, A(\mathbf{R}, l)_{L(n)})| \leq \sum_{|m-l| \leq 1} |(Af, \mathbf{R}, m), A(\mathbf{R}, l)_{L(n)})| \ll$$

$$\ll 2^{-(g(\mathbf{R})-1/p)l} b_n^{-1/p}.$$

Substituting this estimate in (2.4) and performing the summation we find by the remark to Lemma 2.1

$$|\Phi_n(f) - \hat{f}(0)| \ll b_n^{-g(\mathbf{R})}.$$

The upper estimate in Theorem 3 is proved.

We turn to the proof of the lower estimate. Let $N = (N_1, N_2)$ and

$$T(N) = \{f : f(\mathbf{x}) = \sum_{|k_j| \leq N_j, j=1,2} C_{\mathbf{k}} e^{i(\mathbf{k}, \mathbf{x})}\},$$

$$T(N)_\infty = \{f \in T(N) : \| f \|_\infty \leq 1\}.$$

LEMMA 2.2. *Let* $N \geq 1$. *Then* $\delta_{\nu(N)}(T(2N)_\infty) \geq C > 0$, *where* $\nu(N) = N_1 N_2$.

The proof of this lemma is based on the following assertion, which was proved in [11].

THEOREM 2.1. *Let* $\varepsilon > 0$ *and a subspace* $\Psi \subseteq T(N)$ *be such that* $\dim \Psi \geq \varepsilon \dim T(N)$. *Then there exists* t *in* Ψ *such that* $\| t \|_\infty = 1$ *and* $\| t \|_2 \geq C(\varepsilon) > 0$.

Let $\xi^1, \ldots, \xi^m, \xi^j \in \pi_2, j = 1, \ldots, m, m = \nu(N)$ be given. We consider the subspace $\Psi \subseteq T(N)$:

$$\Psi = \{t \in T(N) : t(\xi^j) = 0, \quad j = 1, \ldots, m\}.$$

Then $\dim \Psi \geq C_1 \dim T(N), C_1 > 0$, and by Theorem 2.1 there is $t \in \Psi$ such that

$$\| t \|_\infty = 1 \text{ and } \| t \|_2 \geq C_2 > 0.$$

Setting $f = |t|^2 \in T(2N)_\infty$ we get that for any cubature formula with points $\xi^1, \dots, \xi^m$

$$\Lambda_m(f, \xi) = 0,$$

but

$$\hat{f}(0) = \| t \|_2^2 \geq C_2^2 > 0.$$

The lemma is proved since $\xi^1, \dots, \xi^m$ are arbitrary.

Lemma 2.2 with help of Bernstien's inequality implies the lower estimate in Theorem 3. Here one has to consider

$$N = ([2^{ng(R)/R_1}], [2^{ng(R)/R_2}]).$$

References

[1] N.M. Korobov, *On approximate calculating multiple integrals*, Dokl.Akad.Nauk SSSR **124** no. 6 (1959), pp. 1207–1210.

[2] N.S. Bakhvalov, *On approximate evaluation of multiple integrals*, Vest. Univ. of Moscow **4** (1959), pp. 3–18. (In Russian)

[3] Hua Loo Keng and Wang Yuan, *Remarks concerning numerical integration*, Sci.Rec. New Ser. **4** (1960), pp.,8–11.

[4] N.M. Korobov, *Number-Theoretic methods in approximate analysis*, Fizmat, Moscow, 1963. (In Russian)

[5] V.N. Temlyakov, *On estimates of errors of quadrature formulas for the classes of functions with bounded mixed derivative*, Mat.Zametky **46** no. 2 (1989), pp. 128–134. (In Russian)

[6] V.N. Temlyakov, *Estimates of errors of the Fibonacci quadrature formulas for the classes of functions with bounded mixed derivative*, Trudy Mat. Inst. Steklov **200** (1991), pp. 327–335. (In Russian)

[7] V.N. Temlyakov, *On estimates of errors of cubature formulas* (to appear).

[8] V.N. Temlyakov, *Approximation of Periodic Functions*, Nauka, Moscow, Submitted for publication (In Russian); Nova Science Publishers, New York, Submitted for publication.

[9] V.N. Temlyakov, *Approximation of Functions with Bounded Mixed Derivative*, Proc. Steklov Inst. Math. **178** (1989).

[10] V.N. Temlyakov, *On approach of obtaining lower bounds for the error of quadratures*, Math. USSR Sbornik **181** no. 10 (1990), pp. 1403–1413. (In Russian)

[11] V.N. Temlyakov, *Bilinear approximation and connected questions*, Trudy Mat. Inst. Steklov **194** (1992), pp. 229–249. (In Russian)

MIAN
117966, Moscow
Russia

PARAMETERS OF ORTHOGONAL POLYNOMIALS

S. Khrushchev

1. Introduction

The Herglotz formula

$$\int_{\mathbb{T}} \frac{t+z}{t-z} \, d\sigma(t) = \frac{1+zf}{1-zf}, \quad |z| < 1, \tag{1.1}$$

establishes a one-to-one correspondence between the set of probabilistic measures σ on the unit circle $\mathbb{T}$ and the unit ball $\mathcal{B}$ of the Hardy algebra $H^\infty = \{f : f \text{ is bounded and holomorphic in } \mathbb{D}\}$, $\mathbb{D}$ being the unit disc.

Given a probabilistic measure σ on $\mathbb{T}$ the family of orthogonal polynomials $\{\phi_n\}_{n \geq 0}$ in $L^2(d\sigma)$ is obtained by applying the standard Hilbert-Schmidt orthogonalization process to the sequence $\{z^n\}_{n \geq 0}$. We have

$$\phi_n(z) = k_n z^n + \ldots + l_n, \quad k_n > 0$$

where

$$\|\phi_n\|^2 = \int_{\mathbb{T}} |\phi_n|^2 \, d\sigma = 1.$$

The parameters of $\{\phi_n\}_{n \geq 0}$ are defined by

$$a_n = -\bar{l}_{n+1} \cdot k_{n+1}^{-1}, \quad n = 0, 1, \ldots$$

The parameters of orthogonal polynomials were introduced and studied by Ya.L.Geronimus [2,3].

On the other hand starting with a function f in $\mathcal{B}$ one can define the sequence $\{\gamma_n\}_{n \geq 0}$ of Schur parameters of f by the following algorithm:

$$f_0 \overset{\text{def}}{=} f, \quad f_0(z) = \frac{zf_1 + \gamma_0}{1 + \bar{\gamma}_0 z f_1}, \ldots, \quad f_n(z) = \frac{zf_{n+1} + \gamma_n}{1 + \bar{\gamma}_n z f_{n+1}}, \ldots$$

Since $f_n(0) = \gamma_n$ and $f_n \in B$ the function f_{n+1} can be defined if $|\gamma_n| < 1$. In case $|\gamma_n| = 1$ Schur's algorithm interrupts at the n-th step, $f_n(z) \equiv \gamma_n$ in $\mathbf{D}$ and f turns out to be a Blaschke product of degree n.

The purpose of this paper is to provide a simple proof to the following important theorem due to Ya.L. Geronimus [2].

THEOREM 1. *For any measure* σ

$$a_n = \gamma_n, \ n = 0, 1, \dots$$

2. Proof of Theorem 1

Following [1] we can rewrite Schur's algorithm as follows

$$f(z) = (A_n + zB_n^* f_{n+1})/(B_n + zA_n^* f_{n+1}),$$

A_n, B_n being polynomials in z of degree n. Here we set

$$P_n^*(z) = z^n \cdot \overline{P_n(1/\bar{z})}. \tag{2.1}$$

The adjacent steps of Schur's algorithm in this notation are related by

$$\begin{pmatrix} B_n & A_n \\ A_n^* & B_n^* \end{pmatrix} = \begin{pmatrix} 1 & \gamma_n z \\ \bar{\gamma}_n & z \end{pmatrix} \begin{pmatrix} B_{n-1} & A_{n-1} \\ A_{n-1}^* & B_{n-1}^* \end{pmatrix}. \tag{2.2}$$

LEMMA 1. *For every* n

$$A_n(z) = \gamma_0 + \dots + \gamma_n \cdot z^n$$

$$B_n(z) = 1 + \dots + \gamma_n \bar{\gamma}_0 \cdot z^n$$

Proof. It follows from (2.2) that

$$A_n(z) = A_{n-1} + \gamma_n z B_{n-1}^*$$

$$B_n(z) = B_{n-1} 1 + \gamma_n z A_{n-1}^*.$$

The proof is completed by induction with help of (2.1).

Applying the multiplicative functional det to polynomial matrices in (2.2) we obtain an important identity

$$B_n B_n^* - A_n A_n^* = z^n \cdot \prod_{k=0}^{n}(1 - |\gamma_k|^2) = z^n \cdot \omega_n, \tag{2.3}$$

which restricted to **T** yields

$$|B_n|^2 - |A_n|^2 \equiv \omega_n > 0. \tag{2.4}$$

LEMMA 2. *Suppose that* $|\gamma_n| < 1$. *Then*

$$\inf\{|B_n(z)| : z \in \mathbb{D}\} > 0.$$

Proof. Assuming that $\inf\{|B_{n-1}(z)| : z \in \mathbb{D}\} > 0$ we obtain by (2.4) and by the maximum principle that

$$|B_{n-1}| \geq \max(|A_{n-1}|, |A_{n-1}^*|)$$

in **D**. Now (2.2) implies that

$$|B_n(z)| = |B_{n-1}(z) - \gamma_n z A_{n-1}(z)| \geq |B_{n-1}(z)| - |\gamma_n| \cdot |A_{n-1}(z)| > 0$$

in **D**.

Another consequence of (2.3) and (2.4) is that A_n/B_n is a contractive rational function of degree n holomorphic in **D** and satisfying

$$f(z) - \frac{A_n}{B_n}(z) = \frac{A_n + zB_n^* f_{n+1}}{B_n + zA_n^* f_{n+1}} - \frac{A_n}{B_n} = \frac{z^{n+1}\omega_n f_{n+1}}{B_n(B_n + zA_n^* f_{n+1})} = O(z^{n+1})$$

as $z \to 0$. Substituting this into the Herlotz formula (see (1.1)), we obtain

$$F(z) \overset{\text{def}}{=} \int_{\mathbb{T}} \frac{t+z}{t-z}\, d\sigma(t) = \frac{T_n(z)}{R_n(z)} + O(z^{n+1}), \quad z \to 0, \tag{2.5}$$

where polynomials T_n, R_n are defined by

$$T_n(z) = B_{n-1}(z) + zA_{n-1}(z),$$
$$R_n(z) = B_{n-1} - zA_{n-1}(z). \tag{2.6}$$

It follows from (2.3) that

$$T_n R_n^* + R_n T_n^* = (B_{n-1} + zA_{n-1})(zB_{n-1}^* - A_{n-1}^*)$$
$$+ (B_{n-1} - zA_{n-1})(zB_{n-1}^* + A_{n-1}^*) \tag{2.7}$$
$$= 2z(B_{n-1}B_{n-1}^* - A_{n-1}A_{n-1}^*) = 2z^n \omega_{n-1}.$$

Since $A_{n-1}/B_{n-1} \in \mathcal{B}$ we obtain from (2.6) that R_n does not vanish in $\mathbb{D}$. By (2.7) we have

$$\operatorname{Re} \frac{T_n}{R_n} = \frac{\omega_{n-1}}{|R_n|^2} \qquad \text{on } \mathbb{T}. \tag{2.8}$$

Thus by Herglotz' Theorem, (2.5) and (2.8) we have

$$F(z) = \int_{\mathbb{T}} \frac{t+z}{t-z} \, d\sigma(t) = \int_{\mathbb{T}} \frac{t+z}{t-z} \cdot \frac{\omega_{n-1}}{|R_n|^2} \, dm + O(z^{n+1}), \quad z \to 0,$$

m being the normalized Lebesque measure on $\mathbb{T}$. In other words the Fourier coefficients of $d\sigma$ and $\omega_{n-1} \cdot |R_n|^{-2} dm$ with indices not exceeding n by modulus coincide.

It follows that

$$\int_{\mathbb{T}} P \cdot \bar{t}^k \, d\sigma = \omega_{n-1} \int_{\mathbb{T}} P \bar{t}^k \cdot |R_n|^{-2} \, dm \tag{2.9}$$

for $k = 0, 1, \ldots, n$ and for any polynomial P of degree n. Indeed, both integrals can be expressed as linear combinations of equal Fourier coefficients of $d\sigma$ and $\omega_{n-1} \cdot |R_n|^{-2} \, dm$. Putting now $P = R_n^*$ in (2.9) we obtain

$$\int_{\mathbb{T}} R_n^* \cdot \bar{t}^k \, d\sigma = \omega_{n-1} \int_{\mathbb{T}} \bar{t}^{n-k} \bar{R}_n \cdot |R_n|^{-2} \, dm$$

$$= \omega_{n-1} \int_{\mathbb{T}} \bar{t}^{n-k} R_n^{-1} \, dm = \begin{cases} 0, & k = 0, 1, \ldots, n-1 \\ \omega_{n-1} \cdot R_n^{-1}(0), & k = n. \end{cases}$$

By Lemma 1 we obtain that $R_n(0) = 1$. Now it follows by the definition of orthogonal polynomials that

$$R_n^*(z) = c \cdot \phi_n(z), \quad c > 0. \tag{2.10}$$

Again by Lemma 1 we have

$$R_{n+1}^*(z) = z \cdot B_n^* - A_n^* = z^{n+1} + \ldots - \bar{\gamma}_n$$

which together with (2.10) implies that $k_{n+1} = c^{-1}$, $l_{n+1} = -\bar{\gamma}_n \cdot c^{-1}$, and therefore

$$a_n = -\bar{l}_{n+1} \cdot k_{n+1}^{-1} = \gamma_n \cdot c^{-1} \cdot c = \gamma_n.$$

3. An application

Theorem 1 can be used for study of asymptotic behavior of Schur parameters of contractive holomorphic functions in $\mathbb{D}$.

Suppose that

$$\|f\|_\infty = \sup\{|f(z)| : z \in \mathbb{D}\} < 1$$

and that f belongs to the Hölder class Λ_α, $0 < \alpha < 1$, i.e.

$$|f(z_1) - f(z_2)| \le C_f \cdot |z_1 - z_2|^\alpha, \quad z_1, z_2 \in \mathbb{D}.$$

Taking real parts of the both terms in (1.1) and using the Poisson formula we see that

$$\omega \stackrel{\text{def}}{=} \frac{d\sigma}{dm} = \frac{1 - |f|^2}{|1 - zf|^2} \quad \text{on} \quad \mathbb{T}. \tag{3.1}$$

It follows that $\omega \in \Lambda_\alpha(\mathbb{T})$ and $0 < \inf \omega \le \sup \omega < +\infty$. Let h be an outer function satisfying

$$\omega = |h|^2 \quad \text{on} \quad \mathbb{T}, \quad h(0) > 0.$$

By Szegö theorem [6] we have the following asymptotic formula for orthogonal polynomials on $\mathbb{T}$:

$$\phi_n(z) = \frac{z^n}{\bar{h}} + O(\frac{\log n}{n^\alpha}), \quad |z| = 1.$$

Hence

$$\overline{\phi_{n+1}(0)} = \int_{\mathbb{T}} \overline{\phi_{n+1}} \, dm = \int_{\mathbb{T}} \overline{z^{n+1}} \frac{dm}{h} + O(\frac{\log n}{n^\alpha}). \tag{3.2}$$

The integral in the right hand side of (3.2) is the Fourier coefficient of h^{-1}. Since $h^{-1} \in \Lambda_\alpha$ we obtain that

$$\overline{\phi_{n+1}(0)} = O(\frac{\log n}{n^\alpha}).$$

It is well-known [6] that

$$0 < \lim_n k_n = h(0)^{-1} < +\infty.$$

Therefore

$$\gamma_n = a_n = -\overline{\phi_{n+1}(0)}/k_{n+1} = O(\frac{\log n}{n^\alpha}).$$

This shows that the Schur parameter of contractive holomorphic functions behave very much like their Taylor (Fourier) coefficients. However, there is an important difference.

Let $\mathcal{B}_e$ be the set of all extreme points of $\mathcal{B}$.

THEOREM [1]. *Let $f \in \mathcal{B}$. Then*

$$\sum_n |\gamma_n|^2 < +\infty$$

iff $f \notin \mathcal{B}_e$ or f is a finite Blaschke product.

4. Concluding remarks

The approach presented was developed during author's visit to the Mittag-Leffler Institute in October-November, 1990. The author is indebted to B.L. Golinskii and L.B. Golinskii for interesting correspondence in 1991 concerning asymptotic bahavior of the Schur parameters. Another proof of Theorem 1 can be found in [4]. Asymptotic behavior of the Schur parameters for more delicate classes of functions was studied in [4,5].

References

[1] Boyd D.W., *Schur's algorithm for bounded holomorphic functions*, Bull.London Math. Soc. **11** (1979), pp. 145–150.

[2] Geronimus Ya. L., *Polynomials orthogonal on the circle and their applications*, Zapiski NII matematiki i mehaniki i HMO. Ser. 4, vol. XIX, Kharkov, 1948, pp. 35–120.

[3] Geronimus Ya. L., *Polynomials orthogonal on the circle and the interval*, Fizmatgiz, Moscow, 1958.

[4] Golinskii B. L., *Applications of orthogonal polynomials to the theory of S-functions*, Matematicheskie metody analiza dinamicheskih sistem, N3, Kharkov, 1948, pp. 100–110.

[5] Golinskii B. L., *A relation between the decreasing of orthogonal polynomials parameters and properties of corresponding distributions*, Izv. Akad. Armyanskoi SSR **XV** no. 2 (1980), pp. 127–144, Kharkov.

[6] Grenander and Szegö, *Toeplitz forms and their applications*, University of California Press, Berkeley and Los Angelos, 1958.

EIMI
St.Petersburg
Russia

SOME NUMERICAL RESULTS
ON BEST UNIFORM POLYNOMIAL
APPROXIMATION OF X^α ON $[0,1]$

Amos J. Carpenter*and Richard S. Varga†

Let α be a positive number, and let $E_n(x^\alpha; [0,1])$ denote the error
of best uniform approximation to x^α by polynomials of degree at
most n on the interval $[0,1]$. Russian mathematician S. N. Bernstein
established the existence of a nonnegative constant $\beta(\alpha)$ such that
$\beta(\alpha) := \lim_{n\to\infty} (2n)^{2\alpha} E_n(x^\alpha; [0,1])$ $(\alpha > 0)$.
In addition, Bernstein showed that $\pi\beta(\alpha) < \Gamma(2\alpha)|\sin(\pi\alpha)|$ $(\alpha > 0)$,
and that $\Gamma(2\alpha)|\sin(\pi\alpha)|(1 - 1/(2\alpha - 1)) < \pi\beta(\alpha)$ $(\alpha > 1/2)$, so that
the asymptotic behavior of $\beta(\alpha)$ is thus known when $\alpha \to \infty$.

Still, the problem of trying to determine $\beta(\alpha)$ more precisely, for
all $\alpha > 0$, is intriguing. To this end, we have rigorously determined the
numbers $\{E_n(x^\alpha; [0,1])\}_{n=1}^{40}$ for thirteen values of α, where these numbers were calculated with a precision of at least 200 significant digits.
For each of these thirteen values of α, Richardson's extrapolation was
applied to the products $\{(2n)^{2\alpha} E_n(x^\alpha; [0,1])\}_{n=1}^{40}$ to obtain estimates
of $\beta(\alpha)$ to approximately 40 decimal places. Included are graphs of the
points $(\alpha, \beta(\alpha))$ for the thirteen values of α that we considered.

1. Introduction

One of the first results in *constructive* approximation theory was in 1913
by the great Russian mathematician S. N. Bernstein who studied in [1] the
best uniform approximation by polynomials of the specific function $|t|$ on the
interval $[-1, +1]$. With π_n denoting the set of all real polynomials of degree
at most n $(n = 0, 1, \cdots)$, and with

$$E_n(f(t); [a,b]) := \inf \left\{ \|f(t) - g(t)\|_{L_\infty[a,b]} : g(t) \in \pi_n \right\} \qquad (1.1)$$

*Research was done while a National Science Foundation intern in parallel processing
in the Mathematics and Computer Science Division, Argonne National Laboratory.
†Research supported by the National Science Foundation.
AMS(MOS) subject classification: 41A10, 41A50.
Keywords: Polynomials, best approximation, Remez algorithm.

denoting the error of best uniform approximation of a given continuous function $f(t)$ on the finite interval $[a, b]$ by polynomials in π_n, Bernstein proved in [1] that there exists a positive constant $\beta(\frac{1}{2})$ such that

$$\beta(\tfrac{1}{2}) = \lim_{n \to \infty} 2n E_{2n}(|t|; [-1, +1]). \tag{1.2}$$

In addition, he constructively determined the following upper and lower bounds for $\beta(\frac{1}{2})$:

$$0.278 < \beta(\tfrac{1}{2}) < 0.286. \tag{1.3}$$

Bernstein [1, p. 56] remarked in a footnote that it would be very interesting to determine if the limit of $2n E_{2n}(|t|; [-1, +1])$ is a new transcendental number, or can be expressed in terms of known transcendental numbers. Without resolving this question, he remarked, "as a curious coincidence", that $1/(2\sqrt{\pi}) = 0.28209 \cdots$, this number being nearly the average, namely 0.282, of the upper and lower bounds in (1.3), and this remark became known in the literature as the

Bernstein Conjecture : $\quad \beta(\tfrac{1}{2}) \overset{?}{=} \dfrac{1}{2\sqrt{\pi}} = 0.28209 \cdots . \tag{1.4}$

More than 70 years later, Varga and Carpenter [18] showed in 1985, by means of high-precision computations, that the Bernstein conjecture of (1.4) is *false*, in that $\beta(\frac{1}{2})$ rigorously satisfies

$$0.28016\ 85460 \cdots \leq \beta(\tfrac{1}{2}) \leq 0.28017\ 33791 \cdots . \tag{1.5}$$

In addition, from high-precision extrapolations, the following estimate from [18] for $\beta(\frac{1}{2})$ to 50 significant figures is

$$\beta(\tfrac{1}{2}) = 0.28016\ 94990\ 23869\ 13303\ 64364\ 91230\ 67200\ 00424\ 82139\ 81236. \tag{1.6}$$

But, what is $\beta(\frac{1}{2})$? Knowing $\beta(\frac{1}{2})$ to some 50 significant digits is one thing, but it would be much more mathematically satisfying to have $\beta(\frac{1}{2})$ in *closed form*, i.e., to have $\beta(\frac{1}{2})$ represented as the evaluation of some specific function. The problem of finding such a solution for $\beta(\frac{1}{2})$ has intrigued us for some time.

Indeed, this problem evidently intrigued Bernstein also for, at the end of his long paper [1], Bernstein briefly mentions without proof that for any $\alpha > 0$, it can be similarly shown that there is a nonnegative real number $\delta(\alpha)$ such that*

$$\delta(\alpha) := \lim_{n \to \infty} (2n)^{2\alpha} E_{2n}(x^{\alpha}; [0, 1]) \quad (\alpha > 0). \tag{1.7}$$

---

*We remark that the expression in (1.7) above corrects a typographical error appearing on line 11 of [1, p. 56], i.e., the factor $(2n)^{2\alpha}$ in (1.7) appears incorrectly in [1, p. 56] as $(2n)^{\alpha}$.

It readily follows from (1.7), more generally, that

$$\delta(\alpha) = \lim_{n \to \infty} n^{2\alpha} E_n(x^{\alpha}; [0, 1]) \quad (\alpha > 0). \tag{1.8}$$

Next, to show that the result of (1.8) can be related to the special case $\alpha = \frac{1}{2}$ of (1.2), we note that as $|t|^{2\alpha}$ is an even function on $[-1, +1]$, it is easily seen (cf. Rivlin [15, p. 43, Exercise 1.1] or [17, p. 4]) that

$$E_{2n}(|t|^{2\alpha}; [-1, +1]) = E_n(x^{\alpha}; [0, 1]),$$

which implies from (1.8) that for $\alpha > 0$

$$\beta(\alpha) = \lim_{n \to \infty} (2n)^{2\alpha} E_{2n}(|t|^{2\alpha}; [-1, +1]) = \lim_{n \to \infty} (2n)^{2\alpha} E_n(x^{\alpha}; [0, 1]), \quad (1.9)$$

where (cf. (1.8)) $\beta(\alpha) := 2^{2\alpha} \delta(\alpha)$.

Then, 25 years after the appearance of his paper [1] in 1913, Bernstein revised this problem in 1938 by publishing in [2] or [3] some remarkable results for the function $\beta(\alpha)$. Using a myriad of interesting but difficult techniques, Bernstein established in [2] the upper bound

$$\beta(\alpha) < \frac{\Gamma(2\alpha)|\sin(\pi\alpha)|}{\pi} \quad (\alpha > 0), \tag{1.10}$$

as well as the lower bound

$$\frac{\Gamma(2\alpha)|\sin(\pi\alpha)|}{\pi} \left(1 - \frac{1}{2\alpha - 1}\right) < \beta(\alpha) \quad (\alpha > \frac{1}{2}), \tag{1.11}$$

from which the precise asymptotic behavior as $\alpha \to \infty$, namely,

$$\lim_{\alpha \to \infty} \frac{\beta(\alpha)}{\{\Gamma(2\alpha)|\sin(\pi\alpha)|/\pi\}} = 1, \tag{1.12}$$

follows. He also established in [2] that

$$\lim_{\alpha \to 0} \beta(\alpha) = \frac{1}{2} =: \beta(0). \tag{1.13}$$

The results of (1.10) – (1.13) can be interpreted as an attempt by Bernstein to determine $\beta(\alpha)$ in closed form for $\alpha \geq 0$. But, aside from the specific values $\beta(0)$ and $\beta(\frac{1}{2})$ (cf. (1.13) and (1.5) – (1.6), respectively) and the elementary fact from (1.9) that $\beta(s) = 0$ for any positive integer s, we knew of no other computed or theoretical values of $\beta(\alpha)$. Because of this, it was our thought that similarly determining $\beta(\alpha)$ of (1.9), for a number of different values of α, might shed some light on our problem of determining

$\beta(\frac{1}{2})$ in closed form. To this end, using high-precision calculations as in [18], we present below highly accurate estimates of $\beta(\alpha)$ for the thirteen specific values of α, namely

$$
\begin{cases}
\alpha = \dfrac{j}{8} & (j = 1, 2, \cdots, 7), \\[2mm]
\alpha = \dfrac{j}{4} & (j = 5, 6, 7), \quad \text{and} \\[2mm]
\alpha = \dfrac{j}{4} & (j = 9, 10, 11).
\end{cases}
\tag{1.14}
$$

(All the values of α in (1.14) lie in the interval (0,3), with most of these values lying in (0,1). These choices of α were made to complement the known asymptotic behavior of $\beta(\alpha)$ in (1.12) when $\alpha \to \infty$.) For each α in (1.14), each of the numbers $\{E_n(x^\alpha; [0,1])\}_{n=1}^{40}$ was determined to a precision of at least 200 significant digits. Then for each α in (1.14), the products $\{(2n)^{2\alpha} E_n(x^\alpha; [0,1])\}_{n=1}^{40}$ were extrapolated, using the Richardson extrapolation method, to obtain an estimate of $\beta(\alpha)$ to at least 40 significant digits. As can be really understood, this is a *nontrivial computation*, since for *each* choice of α and n, the number $E_n(x^\alpha; [0,1])$ is determined only after the result of several iterations of the second Remez algorithm (cf. §2). The total amount of these calculations consumed approximately 2200 cpu hours on the Alliant FX/8 and the Encore Multimax at the Argonne National Laboratory.

In Table 1.1, we give our estimates, to 40 significant digits, of $\beta(\alpha)$ which includes the values of α in (1.14), along with $\beta(0) = \frac{1}{2}$ from (1.13) and $\beta(s) = 0$ for s a positive integer.

Table 1.1 Estimates of $\beta(\alpha)$ to 40 decimal places

α	$\beta(\alpha)$
0.000	5.00000 00000 00000 00000 00000 00000 00000 00000E-01
0.125	3.92106 06865 24306 18102 87889 06500 29516 74073E-01
0.250	3.48648 23272 56100 43273 50066 60904 27053 37181E-01
0.375	3.15241 27414 61107 18764 66738 56654 82499 32994E-01
0.500	2.80169 49902 38691 33036 43649 12306 72000 04248E-01
0.625	2.36444 76483 36463 84095 46777 48284 40668 05347E-01
0.750	1.78360 33169 26983 67018 81533 55271 40747 72851E-01
0.875	1.00876 79735 91345 44168 60888 21616 98102 46483E-01
1.000	0.00000 00000 00000 00000 00000 00000 00000 00000E+00
1.250	2.74965 97507 96998 83110 43751 61314 34800 90980E-01
1.500	5.91106 95862 73252 18719 17623 47676 77064 56234E-01
1.750	7.00436 70509 81847 47876 41483 52647 24759 44023E-01
2.000	0.00000 00000 00000 00000 00000 00000 00000 00000E+00
2.250	2.48295 48477 65220 75909 82753 63988 73708 98876E+00
2.500	7.28031 92389 13027 78383 71440 91741 94820 71406E+00
2.750	1.12733 90258 80507 55856 52198 44242 73647 99596E+01
3.000	0.00000 00000 00000 00000 00000 00000 00000 00000E+00

We remark that the behavior of $\beta(\alpha)$ for small positive α particularly interested us, and by the same techniques, we similarly determined the following three values of $\beta(\alpha)$ for α near zero:

$$\begin{cases} \beta(\tfrac{1}{32}) = 0.45292\ 47687\ 14618\ 57962\,, \\ \beta(\tfrac{2}{32}) = 0.42680\ 83946\ 25923\ 80658\,, \\ \beta(\tfrac{3}{32}) = 0.40752\ 60096\ 52622\ 44796\,, \end{cases} \qquad (1.15)$$

but to a lesser accuracy of approximately 20 significant digits (which is surely sufficient for plotting purposes).

In Figure 1.1, we have plotted as crosses, "x", all the values of $\beta(\alpha)$ from Table 1.1 and (1.15), i.e., values of α in the interval $[0, 3]$. In Figure 1.2, we give an enlargement of this plot for α in the interval $[0, 1]$. In addition, Figures 1.1 and 1.2 both contain a plot of the Bernstein function (cf. (1.10) – (1.13)), namely

$$B(\alpha) := \frac{\Gamma(2\alpha)|\sin(\pi\alpha)|}{\pi} \quad (\alpha \geq 0), \qquad (1.16)$$

for purposes of comparison. Unfortunately, while we now know many new highly accurate estimates for $\beta(\alpha)$ of (1.9), we have been unable to find a *closed form expression* for $\beta(\alpha)$ which exactly fits our computed data. In Table 1.2, we give values of the Bernstein function $B(\alpha)$ of (1.16) to ten decimal digits for comparison with our numerical estimates of $\beta(\alpha)$ from Table 1.1 and (1.15).

Table 1.2 Estimates of $\beta(\alpha)$ and values of $(\Gamma(2\alpha)|\sin(\pi\alpha)|)/\pi$

α	$\beta(\alpha)$	$(\Gamma(2\alpha)|\sin(\pi\alpha)|)/\pi$
0.00000	0.50000 00000	0.50000 00000
0.03125	0.45292 47687	0.48301 32572
0.06250	0.42680 83946	0.46785 15819
0.09375	0.40752 60097	0.45415 88569
0.12500	0.39210 60687	0.44164 25034
0.25000	0.34864 82327	0.39894 22804
0.37500	0.31524 12741	0.36037 05302
0.50000	0.28016 94990	0.31830 98862
0.62500	0.23644 47648	0.26655 48303
0.75000	0.17836 03317	0.19947 11402

Table 1.2 (continued)

α	$\beta(\alpha)$	$(\Gamma(2\alpha)|\sin(\pi\alpha)|)/\pi$
0.87500	0.10087 67974	0.11195 27708
1.00000	0.00000 00000	0.00000 00000
1.25000	0.27496 59751	0.29920 67103
1.50000	0.59110 69586	0.63661 97724
1.75000	0.70043 67051	0.74801 67758
2.00000	0.00000 00000	0.00000 00000
2.25000	2.48295 48478	2.61805 87151
2.50000	7.28031 92389	7.63943 72684
2.75000	11.27339 02588	11.78126 42181
3.00000	0.00000 00000	0.00000 00000

It is our fervent hope that our calculations of these values of $\beta(\alpha)$ in Table 1.1 and (1.15) will give insight and will stimulate others to ponder and mathematically solve our question as to what exactly is $\beta(\alpha)$ for all $\alpha \geq 0$.

We mention that this hope, where numerical calculations could stimulate fundamental mathematical research in approximation theory, has had recent precedences in rational approximation theory. Earlier, motivated by a problem in the numerical solution of heat-conduction problems, the "1/9" Conjecture in rational approximation theory arose, and high-precision calculations were carried out on this conjecture. Subsequently, a beautiful mathematical resolution of this conjecture was given recently in 1987 by Gonchar and Rackhmanov [8]. For highlights of this, see [17, Chapter 2]. More recently, high-precision calculations of the best uniform rational approximations $E_{n,n}(|t|; [-1, +1])$ in Varga, Ruttan, and Carpenter [19] gave rise to the

Conjecture : $\lim\limits_{n \to \infty} e^{\pi\sqrt{2n}} \, E_{2n,2n}(|t|; [-1, +1]) \overset{?}{=} 8,$ (1.17)

which was subsequently shown to be mathematically correct in an elegant solution by Stahl [16].

In the next section of this paper, we give in detail a description of how our high-precision calculations of the numbers $E_n(x^\alpha; [0, 1])$ were obtained. Finally, in §3 we discuss our use of the Richardson extrapolation method to accelerate the convergence of the products $\left\{(2n)^{2\alpha} E_n(x^\alpha; [0, 1])\right\}_{n=1}^{40}$. In this final section, we further give numerical evidence for a new conjecture concerning the asymptotic behavior of the products $(2n)^{2\alpha} E_n(x^\alpha; [0, 1])$, as $n \to \infty$. (This conjecture extends the associated conjecture of [18] for the case $\alpha = \frac{1}{2}$ to the general case $\alpha > 0$.)

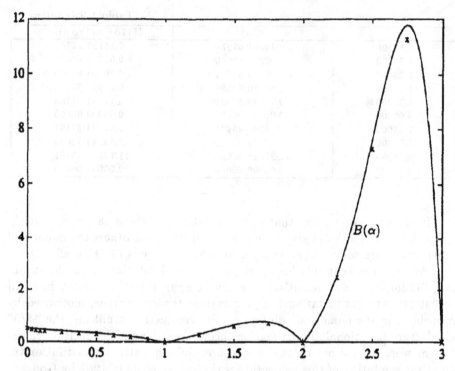

Figure 1.1 The points $(\alpha, \beta(\alpha))$ and the function $B(\alpha)$ for $0 \leq \alpha \leq 3$

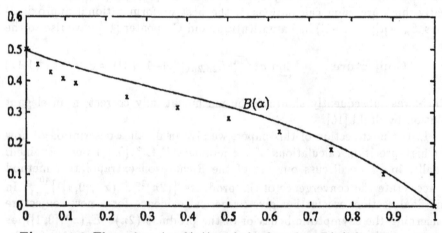

Figure 1.2 The points $(\alpha, \beta(\alpha))$ and the function $B(\alpha)$ for $0 \leq \alpha \leq 1$

2. Computing the Products $\left\{(2n)^{2\alpha}E_n(x^\alpha;[0,1])\right\}_{n=1}^{40}$

For any $\alpha > 0$ not an integer, we consider the best uniform approximation from π_m to x^α on $[0,1]$ for each nonnegative integer m. It is well known (cf. Rivlin [15, p. 28]) that there is a unique p_m^* in π_m such that

$$E_m(x^\alpha;[0,1]) := \|x^\alpha - p_m^*(x)\|_{L_\infty[0,1]}. \qquad (2.1)$$

Moreover, for each nonnegative integer m, it is easily verified that

$$W_m(\alpha) := \operatorname{span}\{1, x, \cdots, x^m; x^\alpha\} \qquad (\alpha > 0 \text{ not an integer}) \qquad (2.2)$$

is a *Haar space* of dimension $m+2$ on the interval $[0,1]$, i.e., any function not identically zero in $W_m(\alpha)$ has at most $m+1$ distinct zeros in $[0,1]$. Thus, in the terminology of Loeb [11], x^α is *hypernormal* on $[0,1]$. Consequently (cf. [11] or Meinardus [13, p. 165]), for any nonnegative integer m, the unique best uniform approximation p_m^* in π_m for which (2.1) is valid, has the property that $\partial p_m^* = m$ (where ∂g denotes the exact degree of a polynomial g), and, moreover, the largest alternation set for $x^\alpha - p_m^*(x)$ on $[0,1]$ consists of $m+2$ points, i.e., there exist $m+2$ points $\{x_j\}_{j=0}^{m+1}$ in $[0,1]$ with $0 \le x_0 < x_1 < \cdots < x_{m+1} = 1$, such that

$$x_i^\alpha - p_m^*(x_i) = \sigma(-1)^i E_m(x^\alpha;[0,1]) \qquad (i = 0,1,\cdots,m+1), \qquad (2.3)$$

with σ satisfying $\sigma = +1$ or $\sigma = -1$.

The minimization problem (2.1) was solved using the following essentially standard implementation of the *second Remez algorithm* (cf. Remez [14] or Meinardus [13, p. 105]):

Step 1: Let $S := \{t_j\}_{j=0}^{m+1}$ be a set of $m+2$ distinct points in $[0,1]$ satisfying

$$0 \le t_0 < t_1 < \cdots < t_{m+1} \le 1. \qquad (2.4)$$

Step 2: Find the unique polynomial $h_m(t) = \sum_{j=0}^{m} a_j t^j$ and a constant λ such that

$$h_m(t_j) + (-1)^j \lambda = t_j^\alpha \qquad (j = 0,1,\cdots,m+1), \qquad (2.5)$$

which is a linear system of $m+2$ equations in the $m+2$ unknowns λ and $\{a_j\}_{j=0}^m$. Because the associated coefficient matrix for (2.5) is a nonsingular Vandermonde matrix of order $m+2$, there are unique values λ and $\{a_j\}_{j=0}^m$ which solve (2.5). Note that since $\alpha > 0$ is not an integer, then $\lambda \ne 0$. From (2.5), $h_m(t)$ is then the best uniform approximation from π_m to x^α on this discrete set S, with an alternating error $|\lambda|$ in successive points t_j of S. Thus, in analogy with (2.1), we can write

$$\|x^\alpha - h_m(x)\|_{L_\infty(S)} =: E_m(x^\alpha; S) = |\lambda| > 0, \qquad (2.6)$$

and as S is a subset of $[0, 1]$, then clearly

$$\|x^\alpha - h_m(x)\|_{L_\infty[0,1]} - |\lambda| \geq 0. \qquad (2.7)$$

Step 3: With a preassigned small $\varepsilon > 0$, if $\|x^\alpha - h_m(x)\|_{L_\infty[0,1]} - |\lambda| \leq \varepsilon$, the iteration is terminated. Otherwise, as $x^\alpha - h_m(x)$ is an element of $W_m(\alpha)$ which takes on, by construction, the value $\lambda \neq 0$ with alternating sign in $m+2$ points of $[0, 1]$, then $x^\alpha - h_m(x)$ has precisely $m + 1$ distinct zeros in $[0, 1]$, and a new set S', from the set of local extrema in $[0, 1]$ (with alternating signs) of $x^\alpha - h_m(x)$, can be determined. With this new set S', **Steps 2** and **3** are repeated until convergence is reached.

Starting with the particular set $S^{(0)} := \left\{ t_j^{(0)} \right\}_{j=0}^{m+1}$, where

$$t_j^{(0)} := \frac{1}{2}\left\{ 1 + \cos\left(\frac{(m+1-j)\pi}{m+1}\right) \right\} \qquad (j = 0, 1, \cdots, m+1) \qquad (2.8)$$

are the $m + 2$ extreme points of the Chebyshev polynomial $T_{m+1}(2t - 1)$ on $[0, 1]$, the above algorithm, with $\varepsilon = 10^{-213}$, was used in conjunction with Brent's multiple-precision (MP) package [5], where the products $\left\{ (2n)^{2\alpha} E_n(x^\alpha; [0, 1]) \right\}_{n=1}^{40}$ were computed with a precision of at least 200 significant digits. Because of the known quadratic convergence of this second Remez algorithm (cf. [13, p. 113]), at most *ten* iterations of the algorithm were needed for convergence in all the cases considered. In Tables 2.2 – 2.14, we list the numbers $\{E_n(x^\alpha; [0, 1])\}_{n=1}^{40}$ and the products $\left\{ (2n)^{2\alpha} E_n(x^\alpha; [0, 1]) \right\}_{n=1}^{40}$, each rounded to 25 decimal digits, for $\alpha = \frac{j}{8}$ ($j = 1, 2, \cdots, 7$), $\alpha = \frac{j}{4}$ ($j = 5, 6, 7$), and $\alpha = \frac{j}{4}$ ($j = 9, 10, 11$).

These computations were performed at the Argonne National Laboratory using the **p4** package and up to eight processors on the Alliant FX/8 and the Encore Multimax. Table 2.1 gives the speedup and efficiency on the Encore Multimax for $\alpha = \frac{1}{2}$ and $n = 10$. For a FORTRAN tutorial on the **p4** package, see Lusk and Overbeek [12].

Table 2.1 Best uniform polynomial approximation of $x^{\frac{1}{2}}$ on $[0,1]$

The degree of the polynomial is 10		
Number of Processors	Speedup	Efficiency
1	1.00	1.00
2	1.37	0.69
3	2.15	0.72
4	2.96	0.74
5	3.85	0.77
6	4.75	0.79
7	5.69	0.81
8	6.70	0.84

Figure 2.1 gives graphs of the error functions $E^{(j)}(x) = x^{\frac{1}{2}} - p_n^{(j)}(x)$, for $j = 1$ and 9, and $n = 5$, where $E^{(j)}(x)$ is the error function at the jth iteration of the second Remez algorithm (cf. **Step 3** above). For $n = 5$, *nine* iterations of the Remez algorithm were needed for convergence. The error function $E^{(1)}(x) = x^{\frac{1}{2}} - p_n^{(1)}(x)$ is denoted by a dashed line, while the error function $E^{(9)}(x) = x^{\frac{1}{2}} - p_n^{(9)}(x)$ is denoted by a solid line. In the graph of $E^{(9)}(x)$, there are seven alternation points (denoted by small dark disks) in the interval $[0,1]$.

Figure 2.2 gives graphs of the zeros of the polynomials of best uniform approximation to x^α on $[0,1]$, for $n = 30$. The innermost zeros correspond to the case $\alpha = \frac{5}{4}$, the middle zeros correspond to the case $\alpha = \frac{6}{4}$, and the outermost zeros correspond to the case $\alpha = \frac{7}{4}$. In each of these three cases, there are two zeros on the real axis (one on either side of the origin). These zeros were computed with a precision of at least 200 significant digits by using Jenkins's algorithm [10] and Brent's MP package [5].

It follows from a general result of Blatt and Saff [4, Cor. 2.2] that every point of $[0, 1]$ is a limit point of the zeros of the polynomials of best uniform approximation to x^α on $[0, 1]$. It also follows from their result [4, Thm. 3.4] that, for n large, these zeros are roughly uniformly distributed in angle, which is quite evident from Figure 2.2. Moreover, Figure 2.2 suggests a *further* uniformity in α as well!

The high-precision coefficients of the polynomials of best uniform approximation to x^α on $[0,1]$, together with the extreme points on $[0,1]$ of these polynomials, are much too lengthy to reproduce here. These numbers are, however, available upon request.

Table 2.2 The numbers $\left\{ E_n(x^{\frac{1}{8}};[0,1]) \right\}_{n=1}^{40}$

and the products $\left\{ (2n)^{\frac{1}{4}} E_n(x^{\frac{1}{8}};[0,1]) \right\}_{n=1}^{40}$

n	$E_n(x^{\frac{1}{8}};[0,1])$	$(2n)^{\frac{1}{4}} E_n(x^{\frac{1}{8}};[0,1])$
1	3.25061 25074 87074 67924 96791E-01	3.86565 15220 20465 11260 56734E-01
2	2.76150 31636 19477 56038 05105E-01	3.90535 52265 26873 33084 14186E-01
3	2.50074 11823 92040 64282 53365E-01	3.91387 14633 16022 93554 90478E-01
4	2.32904 52719 18662 58415 35389E-01	3.91697 16402 40032 39094 59871E-01
5	2.20349 50093 39596 87398 41689E-01	3.91842 98052 32129 01695 61450E-01
6	2.10574 24648 79833 28223 10718E-01	3.91922 83396 69610 32938 37563E-01
7	2.02638 59148 93118 78035 43392E-01	3.91971 20766 95181 38166 57091E-01
8	1.96001 34789 36635 49477 29651E-01	3.92002 69578 73270 98954 59302E-01
9	1.90324 58460 34424 28397 31860E-01	3.92024 32604 39412 08769 73143E-01
10	1.85384 18469 46552 96105 35779E-01	3.92039 81918 49731 19796 56528E-01
11	1.81024 44353 79056 85532 88508E-01	3.92051 29375 97544 03801 84001E-01
12	1.77133 11325 06794 29013 58591E-01	3.92060 02762 95127 64989 24387E-01
13	1.73626 80061 27020 41260 05362E-01	3.92066 82853 84276 49900 86341E-01
14	1.70441 98051 20307 84723 57395E-01	3.92072 22728 79461 09515 37814E-01
15	1.67529 22998 24192 29229 64142E-01	3.92076 58429 72185 54474 78658E-01
16	1.64849 39683 37033 18169 42066E-01	3.92080 15123 70940 47946 61591E-01
17	1.62370 97668 10908 85046 05285E-01	3.92083 10814 45673 43683 99437E-01
18	1.60068 27042 91669 04868 12329E-01	3.92085 58656 12882 34696 80135E-01
19	1.57920 06061 87634 51593 37378E-01	3.92087 68440 13578 88385 41443E-01
20	1.55908 64156 54444 16094 73875E-01	3.92089 47576 16884 63653 35012E-01
21	1.54019 09617 98407 29980 10553E-01	3.92091 01754 92631 59342 02779E-01
22	1.52238 74823 43275 89005 41781E-01	3.92092 35405 29909 55585 17763E-01
23	1.50556 74164 70474 85826 99317E-01	3.92093 52015 75815 53553 16537E-01
24	1.48963 71320 40592 04078 26812E-01	3.92094 54364 15407 70318 63163E-01
25	1.47451 53501 93545 65626 44243E-01	3.92095 44684 58368 44490 94276E-01
26	1.46013 10973 33818 04292 04726E-01	3.92096 24790 27471 16049 00523E-01
27	1.44642 20607 67641 61236 88881E-01	3.92096 96165 25365 09135 38932E-01
28	1.43333 32567 13289 83584 80543E-01	3.92097 60033 54063 41282 04326E-01
29	1.42081 59425 11973 84201 14453E-01	3.92098 17411 95359 93257 29494E-01
30	1.40882 67215 42159 54512 14575E-01	3.92098 69150 81299 77891 52150E-01
31	1.39732 68015 25676 72413 95525E-01	3.92099 15965 61461 87384 36639E-01
32	1.38628 13759 22269 79117 85810E-01	3.92099 58461 89017 82822 04973E-01
33	1.37565 91048 40609 74168 18212E-01	3.92099 97154 88001 60397 77349E-01
34	1.36543 16769 74304 85117 28928E-01	3.92100 32485 21918 12990 05600E-01
35	1.35557 34379 41934 60828 61825E-01	3.92100 64831 53409 99243 44816E-01
36	1.34606 10733 84635 91321 14816E-01	3.92100 94520 62610 25619 46486E-01
37	1.33687 33374 79490 36797 19733E-01	3.92101 21835 75600 51505 97090E-01
38	1.32799 08193 26498 45605 62452E-01	3.92101 47023 42388 47129 31085E-01
39	1.31939 57410 82219 26018 54961E-01	3.92101 70298 94849 65112 86498E-01
40	1.31107 17828 33869 87246 95589E-01	3.92101 91851 08320 09918 77773E-01

Table 2.3 The numbers $\left\{E_n(x^{\frac{1}{4}};[0,1])\right\}_{n=1}^{40}$ and the products $\left\{(2n)^{\frac{1}{2}}E_n(x^{\frac{1}{4}};[0,1])\right\}_{n=1}^{40}$

n	$E_n(x^{\frac{1}{4}};[0,1])$	$(2n)^{\frac{1}{2}}E_n(x^{\frac{1}{4}};[0,1])$
1	2.36235 19685 52887 18393 85199E-01	3.34087 01930 26272 39277 58483E-01
2	1.72155 20693 53434 23175 24514E-01	3.44310 41387 06868 46350 49028E-01
3	1.41513 44484 23286 49573 64690E-01	3.46635 73160 71970 65124 32197E-01
4	1.22858 91806 58371 69612 20483E-01	3.47497 49637 43835 76361 54110E-01
5	1.10017 51759 47278 04268 06995E-01	3.47905 93811 69893 45338 54704E-01
6	1.00496 61792 14542 40903 35165E-01	3.48130 49645 75914 49164 08276E-01
7	9.30782 28589 35067 22815 22752E-02	3.48266 84154 91773 97088 34803E-01
8	8.70889 30422 98988 17493 53323E-02	3.48355 72169 19595 26997 41329E-01
9	8.21226 35742 02892 54425 11162E-02	3.48416 83573 26083 65640 94517E-01
10	7.79181 67820 41235 08955 10075E-02	3.48460 63985 73572 84424 92057E-01
11	7.42989 78211 46256 18165 23723E-02	3.48493 09834 18216 08916 08061E-01
12	7.11409 00748 72941 12346 19491E-02	3.48517 81335 27376 15410 66328E-01
13	6.83537 41951 97706 11255 30736E-02	3.48537 06404 02168 04912 27078E-01
14	6.58702 02497 27380 98911 01097E-02	3.48552 34923 45045 03442 21083E-01
15	6.36389 13984 71150 98536 00047E-02	3.48564 68724 55746 99266 23958E-01
16	6.16198 99344 58244 29689 13250E-02	3.48574 78946 06939 25707 40318E-01
17	5.97815 19597 52110 80664 19274E-02	3.48583 16497 38971 69450 22314E-01
18	5.80983 64310 48405 21159 11771E-02	3.48590 18586 29043 12695 47063E-01
19	5.65497 59472 69631 60724 89992E-02	3.48596 12915 80166 92687 43042E-01
20	5.51186 90072 56646 44583 46393E-02	3.48601 20454 84431 15273 77117E-01
21	5.37910 07745 81170 01589 62851E-02	3.48605 57310 66543 37107 86252E-01
22	5.25548 38464 65544 14357 71552E-02	3.48609 36020 99995 15069 74266E-01
23	5.14001 33212 96438 48956 92524E-02	3.48612 66462 69212 71024 46608E-01
24	5.03183 22576 34555 49267 50208E-02	3.48615 56501 54823 54735 25193E-01
25	4.93020 47995 95714 55809 92872E-02	3.48618 12464 32593 32272 61660E-01
26	4.83449 50359 77580 84248 23412E-02	3.48620 39486 38658 28247 18803E-01
27	4.74415 02015 93314 12358 51536E-02	3.48622 41771 07670 86018 55003E-01
28	4.65868 72048 71055 97265 86897E-02	3.48624 22785 55006 51761 13974E-01
29	4.57768 17304 83914 70678 70822E-02	3.48625 85410 22395 34351 48942E-01
30	4.50075 93549 48194 16810 85234E-02	3.48627 32054 00318 43844 06859E-01
31	4.42758 82500 71716 97980 37970E-02	3.48628 64743 94687 48956 53421E-01
32	4.35787 31494 57048 15726 64285E-02	3.48629 85195 65638 52581 31428E-01
33	4.29135 03275 76624 28860 84541E-02	3.48630 94868 97960 46745 19530E-01
34	4.22778 33965 51692 37086 85511E-02	3.48631 95012 43044 68696 75310E-01
35	4.16695 97677 90752 24531 10308E-02	3.48632 86698 86236 05750 73279E-01
36	4.10868 76576 83979 00547 32782E-02	3.48633 70854 30976 58509 08495E-01
37	4.05279 35411 86147 48926 16554E-02	3.48634 48281 45275 92024 10976E-01
38	3.99911 99762 13718 23753 60697E-02	3.48635 19678 82075 99421 14286E-01
39	3.94752 37366 89766 12357 92298E-02	3.48635 85656 59694 80740 99703E-01
40	3.89787 42037 92779 51632 01329E-02	3.48636 46749 69409 43408 78964E-01

Table 2.4 The numbers $\left\{E_n(x^{\frac{3}{8}}; [0,1])\right\}_{n=1}^{40}$

and the products $\left\{(2n)^{\frac{3}{4}} E_n(x^{\frac{3}{8}}; [0,1])\right\}_{n=1}^{40}$

n	$E_n(x^{\frac{3}{8}}; [0,1])$	$(2n)^{\frac{3}{4}} E_n(x^{\frac{3}{8}}; [0,1])$
1	1.73487 73710 33531 31898 95805E-01	2.91770 43244 13769 90067 07448E-01
2	1.08881 44520 03970 20386 43065E-01	3.07963 23298 63688 04251 04245E-01
3	8.13387 98719 42837 00491 31229E-02	3.11825 18729 67257 71251 51648E-01
4	6.58587 41685 17873 45080 94852E-02	3.13278 73678 85633 88615 11356E-01
5	5.58330 58608 08861 98230 36238E-02	3.13972 36167 10298 15503 94538F-01
6	4.87567 25909 44760 10659 95575E-02	3.14355 06981 87253 71433 22908E-01
7	4.34656 34054 60756 11306 08698E-02	3.14587 92296 34213 90729 43537E-01
8	3.93424 89374 66344 59773 81070E-02	3.14739 91499 73075 67819 04856E-01
9	3.60281 32950 00570 46976 22223E-02	3.14844 51774 96947 92650 95306E-01
10	3.32986 87171 91745 48399 35790E-02	3.14919 53983 66262 37691 55305E-01
11	3.10069 67146 81401 31791 30410E-02	3.14975 15601 56162 92033 11986E-01
12	2.90520 16026 52247 84958 55395E-02	3.15017 51868 56524 14302 85155E-01
13	2.73621 48484 90751 32749 41307E-02	3.15050 52402 13713 57138 16117E-01
14	2.58849 78839 22525 34150 36432E-02	3.15076 73599 54383 00886 97099E-01
15	2.45812 83991 95828 19616 44130E-02	3.15097 89750 89857 33985 93379E-01
16	2.34210 79349 74708 68142 60784E-02	3.15115 22666 52019 98453 15962E-01
17	2.23810 26641 29726 46673 65386E-02	3.15129 59548 39322 48172 05332E-01
18	2.14426 72730 91245 86035 06546E-02	3.15141 64147 33597 51793 71027E-01
19	2.05912 23323 27295 34346 58932E-02	3.15151 83940 25613 58524 20920E-01
20	1.98146 70482 91600 66388 38797E-02	3.15160 54870 45516 06732 87775E-01
21	1.91031 59891 58151 44085 04647E-02	3.15168 04552 41475 34662 67740E-01
22	1.84485 24059 86596 15819 66677E-02	3.15174 54484 12208 63521 47539E-01
23	1.78439 32638 73782 49827 10965E-02	3.15180 21603 16007 96340 40995E-01
24	1.72836 26795 39408 12262 51961E-02	3.15185 19399 89879 75141 45284E-01
25	1.67627 14886 93975 03715 05173E-02	3.15189 58726 17548 06511 34242E-01
26	1.62770 13470 06656 89088 35612E-02	3.15193 48391 04983 89107 58458E-01
27	1.58229 22278 98859 95565 57115E-02	3.15196 95605 36416 35540 38997E-01
28	1.53973 24958 81005 37470 04776E-02	3.15200 06317 41446 50422 34577E-01
29	1.49975 09542 17667 65501 35812E-02	3.15202 85469 17496 08886 56083E-01
30	1.46211 04214 31145 49887 30691E-02	3.15205 37193 85731 35947 21098E-01
31	1.42660 25028 17463 18435 19291E-02	3.15207 64969 66566 24131 73293E-01
32	1.39304 33042 06255 80971 85582E-02	3.15209 71740 50425 87345 75602E-01
33	1.36126 98947 09746 23545 92522E-02	3.15211 60011 51194 72009 22680E-01
34	1.33113 73693 67608 15700 00541E-02	3.15213 31925 24846 89698 94543E-01
35	1.30251 63956 89931 00033 17824E-02	3.15214 89322 88411 11807 90339E-01
36	1.27529 11531 33332 51083 98288E-02	3.15216 33793 67350 86333 85028E-01
37	1.24935 75936 44666 35272 14323E-02	3.15217 66715 20857 58198 17314E-01
38	1.22462 19660 98375 77508 95931E-02	3.15218 89286 36340 20556 36175E-01
39	1.20099 95588 41620 84669 59248E-02	3.15220 02554 40888 27403 64255E-01
40	1.17841 36234 52769 31538 97987E-02	3.15221 07437 44701 98776 15305E-01

Table 2.5 The numbers $\left\{E_n(x^{\frac{1}{2}};[0,1])\right\}_{n=1}^{40}$
and the products $\left\{2nE_n(x^{\frac{1}{2}};[0,1])\right\}_{n=1}^{40}$

n	$E_n(x^{\frac{1}{2}};[0,1])$	$2nE_n(x^{\frac{1}{2}};[0,1])$
1	1.25000 00000 00000 00000 00000E-01	2.50000 00000 00000 00000 00000E-01
2	6.76208 99277 78427 52693 01688E-02	2.70483 59711 11371 01077 20675E-01
3	4.59290 62066 86256 43414 24906E-02	2.75574 37240 11753 86048 54943E-01
4	3.46897 28084 38158 70584 45601E-02	2.77517 82467 50526 96467 56481E-01
5	2.78451 18553 55086 01522 28751E-02	2.78451 18553 55086 01522 28751E-01
6	2.32473 26455 41322 55307 47788E-02	2.78967 91746 49587 06368 97346E-01
7	1.99487 81782 75128 74717 46620E-02	2.79282 94495 85180 24604 45268E-01
8	1.74680 52349 65671 54823 39079E-02	2.79488 83759 45074 47717 42527E-01
9	1.55350 36522 78490 00696 06363E-02	2.79630 65741 01282 01252 91453E-01
10	1.39866 21688 59869 14844 94695E-02	2.79732 43377 19738 29689 89390E-01
11	1.27185 41694 94290 33563 40915E-02	2.79807 91728 87438 73839 50014E-01
12	1.16610 59671 82471 96998 22401E-02	2.79865 43212 37932 72795 73762E-01
13	1.07657 79012 13683 42447 86005E-02	2.79910 25431 55576 90364 43613E-01
14	9.99806 63637 77933 30272 38241E-03	2.79945 85818 57821 32476 26708E-01
15	9.33248 68895 46916 41050 78704E-03	2.79974 60668 64074 92315 23611E-01
16	8.74994 22486 34897 75855 32187E-03	2.79998 15195 63167 28273 70300E-01
17	8.23581 40333 32272 17056 86402E-03	2.80017 67713 32972 53799 33377E-01
18	7.77872 35393 05375 26799 72603E-03	2.80034 04741 49935 09647 90137E-01
19	7.36968 17706 81732 93581 51537E-03	2.80047 90728 59058 51560 97584E-01
20	7.00149 36190 10578 81636 37648E-03	2.80059 74476 04231 52654 55059E-01
21	6.66833 17817 09748 34944 82384E-03	2.80069 93483 18094 30676 82601E-01
22	6.36542 65789 83807 59630 73182E-03	2.80078 76947 52875 34237 52200E-01
23	6.08883 64947 19034 14122 99798E-03	2.80086 47875 70755 70496 57907E-01
24	5.83527 59570 58510 53066 43695E-03	2.80093 24593 88085 05471 88973E-01
25	5.60198 43690 47656 71171 20585E-03	2.80099 21845 23828 35585 60292E-01
26	5.38662 53074 33782 78643 85926E-03	2.80104 51598 65567 04894 80682E-01
27	5.18720 80837 44558 98625 33989E-03	2.80109 23652 22061 85257 68354E-01
28	5.00202 60873 20540 78300 00200E-03	2.80113 46088 99502 83848 00112E-01
29	4.82960 78663 70683 06763 25939E-03	2.80117 25624 94996 17922 69045E-01
30	4.66867 79795 44380 47231 57306E-03	2.80120 67877 26628 28338 94384E-01
31	4.51812 54150 26788 45969 04817E-03	2.80123 77573 16608 84500 80987E-01
32	4.37697 79240 42686 22563 12526E-03	2.80126 58713 87319 18440 40017E-01
33	4.24438 10158 16735 10914 80065E-03	2.80129 14704 39045 17203 76843E-01
34	4.11958 06554 41361 92204 59561E-03	2.80131 48457 00126 10699 12501E-01
35	4.00190 89249 14714 95376 79571E-03	2.80133 62474 40300 46763 75700E-01
36	3.89077 20717 95432 18239 95251E-03	2.80135 58916 92711 17132 76581E-01
37	3.78564 04942 20766 17066 80897E-03	2.80137 39657 23366 96629 43864E-01
38	3.68604 03059 31353 89360 76349E-03	2.80139 06325 07828 95914 18025E-01
39	3.59154 61979 69006 18185 45295E-03	2.80140 60344 15824 82184 65330E-01
40	3.50177 53703 24974 26089 62568E-03	2.80142 02962 59979 40871 70054E-01

Table 2.6 The numbers $\left\{ E_n(x^{\frac{5}{8}};[0,1]) \right\}_{n=1}^{40}$

and the products $\left\{ (2n)^{\frac{3}{4}} E_n(x^{\frac{5}{8}};[0,1]) \right\}_{n=1}^{40}$

n	$E_n(x^{\frac{5}{8}};[0,1])$	$(2n)^{\frac{3}{4}} E_n(x^{\frac{5}{8}};[0,1])$
1	8.56645 82174 20506 05128 94161E-02	2.03745 86125 05998 52948 72411E-01
2	3.98832 53055 92170 97502 68410E-02	2.205613 74953 29706 82337 30572E-01
3	2.46263 41002 85478 13783 76902E-02	2.31253 83940 31673 16583 77529E-01
4	1.73502 52197 58018 34401 90662E-02	2.33436 23802 70889 47802 92050E-01
5	1.31863 92765 34615 67886 74134E-02	2.34490 90747 30128 34844 16834E-01
6	1.05252 67139 74341 96148 38477E-02	2.35076 75384 62292 12653 47809E-01
7	8.69381 29915 50372 98089 22437E-03	2.35434 62740 77294 78915 66548E-01
8	7.36465 06443 14552 33237 22986E-03	2.35668 82061 80656 74635 91355E-01
9	6.36075 87259 04069 21183 82768E-03	2.35830 27301 08786 34440 52216E-01
10	5.57860 36894 61179 51118 48636E-03	2.35946 20925 43903 91675 74542E-01
11	4.95385 12759 79454 92711 65022E-03	2.36032 23302 58038 14733 18128E-01
12	4.44455 09614 20568 78709 61656E-03	2.36097 80112 91596 01834 51965E-01
13	4.02225 18978 42024 53212 86014E-03	2.36148 91266 62504 25682 99244E-01
14	3.66701 83712 29253 74117 10891E-03	2.36189 52072 63284 07997 01335E-01
15	3.36449 08647 19817 63888 39483E-03	2.36222 31528 37818 82273 40345E-01
16	3.10407 94814 54856 29725 78162E-03	2.36249 17791 23246 12311 48873E-01
17	2.87781 32791 91449 45571 71183E-03	2.36271 45651 13584 60443 23213E-01
18	2.67958 45694 17517 06371 10509E-03	2.36290 13704 01437 77671 59948E-01
19	2.50463 91730 44923 50997 25521E-03	2.36305 95410 96103 89683 90530E-01
20	2.34922 44053 45154 27910 15228E-03	2.36319 46408 86409 44175 92829E-01
21	2.21034 08668 77184 70415 92540E-03	2.36331 09456 59041 25822 65423E-01
22	2.08556 41419 42231 87308 85484E-03	2.36341 17851 58317 21562 98934E-01
23	1.97291 45570 53180 06692 96416E-03	2.36349 97834 30203 84612 63969E-01
24	1.87076 05977 83603 10939 84636E-03	2.36357 70309 10531 40869 27181E-01
25	1.77774 62971 52898 60588 85711E-03	2.36364 52094 89351 78839 79550E-01
26	1.69273 59592 11511 88182 04100E-03	2.36370 56846 78522 83563 54441E-01
27	1.61477 15947 64033 70304 14781E-03	2.36375 95744 09693 35083 63154E-01
28	1.54303 97990 68065 04878 97997E-03	2.36380 78009 95422 39725 50132E-01
29	1.47684 57256 33846 19162 91417E-03	2.36385 11308 01817 41360 94515E-01
30	1.41559 24513 83066 47608 71837E-03	2.36389 02048 44370 74856 64633E-01
31	1.35876 44794 28844 86316 22399E-03	2.36392 55626 04520 29963 70763E-01
32	1.30591 44471 40939 97463 61801E-03	2.36395 76607 30469 95448 99856E-01
33	1.25665 23389 80074 61157 73860E-03	2.36398 68878 40364 57821 50013E-01
34	1.21063 66726 77617 20900 61703E-03	2.36401 35763 29137 58054 38572E-01
35	1.16756 72519 72893 22062 43157E-03	2.36403 80118 52513 54292 41062E-01
36	1.12717 91718 92179 95286 11647E-03	2.36406 04409 96041 36392 26436E-01
37	1.08923 78322 44156 70669 48295E-03	2.36408 10775 15459 97806 01173F-01
38	1.05353 47678 12082 77646 89617E-03	2.36410 01074 44616 15116 52465E-01
39	1.01988 41440 60420 A6095 20144E-03	2.36411 76932 99817 78260 65131E-01
40	9.88119 79822 36023 15155 74049E-04	2.36413 39775 58755 33541 41055E-01

Table 2.7 The numbers $\left\{ E_n(x^{\frac{3}{4}};[0,1]) \right\}_{n=1}^{40}$
and the products $\left\{ (2n)^{\frac{3}{2}} E_n(x^{\frac{3}{4}};[0,1]) \right\}_{n=1}^{40}$

n	$E_n(x^{\frac{3}{4}};[0,1])$	$(2n)^{\frac{3}{2}} E_n(x^{\frac{3}{4}};[0,1])$
1	5.27343 75000 00000 00000 00000E-02	1.49155 33665 65373 68428 30311E-01
2	2.10499 41369 32078 98811 84740E-02	1.68399 53095 45663 19049 47792E-01
3	1.18077 79420 13000 16429 15090E-02	1.73538 20744 79284 49365 72463E-01
4	7.75843 03959 08188 44049 57467E-03	1.75553 23981 79363 41865 38537E-01
5	5.58246 91986 80911 02507 98277E-03	1.76533 17635 56671 88376 35929E-01
6	4.25986 76577 44793 48453 22760E-03	1.77079 37320 16018 72132 53034E-01
7	3.38684 34457 92223 03980 17226E-03	1.77413 70915 51353 23743 95222E-01
8	2.77551 22996 51750 31034 51760E-03	1.77632 78717 77120 19862 09126E-01
9	2.32800 44363 25493 47884 23229E-03	1.77783 95414 42772 43656 75017E-01
10	1.98889 94337 68184 66050 65861E-03	1.77892 57337 26600 64198 91089E-01
11	1.72472 79498 24355 97384 23183E-03	1.77973 20545 98104 75202 54440E-01
12	1.51421 57504 63441 42391 29027E-03	1.78034 68555 78047 42028 84127E-01
13	1.34326 44555 21703 26451 44801E-03	1.78082 62343 61395 60650 31758E-01
14	1.20220 18461 12590 06908 85230E-03	1.78120 71818 89327 55938 18876E-01
15	1.08419 54311 90818 81842 09057E-03	1.78151 48832 21754 89105 50599E-01
16	9.84296 48562 36347 49176 78151E-04	1.78176 69623 87351 59725 99724E-01
17	8.98842 26819 74662 20507 48822E-04	1.78197 60491 50441 56178 61438E-01
18	8.25070 08546 65244 31477 87944E-04	1.78215 13846 07692 77199 22196E-01
19	7.60861 09335 82573 42912 29810E-04	1.78229 98557 20532 35437 05871E-01
20	7.04566 00868 58775 27745 71266E-04	1.78242 66795 05080 65390 23616E-01
21	6.54883 94183 21856 18638 54082E-04	1.78253 58661 17556 67912 45245E-01
22	6.10776 51210 28255 33349 73375E-04	1.78263 05388 38191 74441 33492E-01
23	5.71405 90320 40865 20633 30799E-04	1.78271 31593 24491 33633 46630E-01
24	5.36089 47779 92470 80129 49516E-04	1.78278 56888 66618 15107 13538E-01
25	5.04266 04666 75800 20723 29464E-04	1.78284 97056 03889 34079 39993E-01
26	4.75470 47133 03977 83535 03120E-04	1.78290 64909 24811 07787 65024E-01
27	4.49314 31208 87924 89774 44627E-04	1.78295 70939 70426 61899 19414E-01
28	4.25470 92009 59948 64448 81724E-04	1.78300 23803 58846 71504 85932E-01
29	4.03663 83633 97386 98970 37043E-04	1.78304 30693 91498 65232 55660E-01
30	3.83657 67922 72660 60362 75536E-04	1.78307 97627 50146 41803 56890E-01
31	3.65250 92458 72075 08152 29990E-04	1.78311 29668 37654 98492 85112E-01
32	3.48270 13873 46932 78007 09207E-04	1.78314 31103 21629 58339 63114E-01
33	3.32565 33735 98010 00009 51936E-04	1.78317 05580 22746 85672 11694E-01
34	3.18006 22407 50505 46984 17017E-04	1.78319 56219 92768 87502 84756E-01
35	3.04479 12172 21207 53491 08453E-04	1.78321 85704 13710 97720 85119E-01
36	2.91884 45329 71644 25701 85133E-04	1.78323 96347 94410 74649 74723E-01
37	2.80134 66197 09326 97390 22336E-04	1.78325 90158 26779 70296 82413E-01
38	2.69152 48420 79778 57294 92670E-04	1.78327 68881 79563 80617 18026E-01
39	2.58869 50858 89504 45744 18379E-04	1.78329 34044 44302 12511 91127E-01
40	2.49224 96715 56483 10256 60075E-04	1.78330 86984 00582 50211 60107E-01

Table 2.8 The numbers $\left\{ E_n(x^{\frac{7}{8}}; [0,1]) \right\}_{n=1}^{40}$

and the products $\left\{ (2n)^{\frac{7}{4}} E_n(x^{\frac{7}{8}}; [0,1]) \right\}_{n=1}^{40}$

n	$E_n(x^{\frac{7}{8}}; [0,1])$	$(2n)^{\frac{7}{4}} E_n(x^{\frac{7}{8}}; [0,1])$
1	2.45434 93986 12976 07421 87500E-02	8.25541 44442 95046 75759 65186E-02
2	8.34746 06218 59568 01190 45596E-03	9.44407 36182 47320 77598 37409E-02
3	4.24868 26581 56727 15605 72621E-03	9.77279 93516 15869 55371 29480E-02
4	2.60244 74754 05414 79124 92647E-03	9.90351 69733 53562 27996 55449E-02
5	1.77250 26925 20828 17196 23793E-03	9.96751 51301 56243 61837 56651E-02
6	1.29293 55897 52071 59011 38915E-03	1.00033 17901 43122 46386 00750E-01
7	9.89401 42657 89072 20978 67617E-04	1.00252 81930 15314 32909 85150E-01
8	7.84351 15022 72376 70692 63043E-04	1.00396 94722 90864 21848 65669E-01
9	6.38886 96639 12825 51076 47339E-04	1.00496 49433 58227 90692 05234E-01
10	5.31688 94761 54876 79930 10117E-04	1.00568 07216 14153 15313 79459E-01
11	4.50246 08769 24362 39007 98476E-04	1.00621 23402 74013 54475 01812E-01
12	3.86807 55787 32884 19431 69683E-04	1.00661 78428 00515 14724 90342E-01
13	3.36354 85280 67543 89731 66182E-04	1.00693 41192 27184 66584 30052F-01
14	2.95517 30374 72720 14164 68463E-04	1.00718 55132 85200 20865 41768E-01
15	2.61959 91453 85459 87585 57004E-04	1.00738 86091 10897 65694 13542E-01
16	2.34021 79600 90179 93213 40468E-04	1.00755 50174 98722 66853 06756E-01
17	2.10494 21454 72886 48282 56809E-04	1.00769 30622 23668 20360 23636E-01
18	1.90479 73809 13474 12252 45572E-04	1.00780 88356 71608 48127 53286E-01
19	1.73300 22053 95899 31361 27938E-04	1.00790 68794 26300 44985 39720E-01
20	1.58435 14009 49230 21957 47720E-04	1.00799 06346 04793 60298 24680E-01
21	1.45479 33506 59923 61165 20638E-04	1.00806 27467 91170 75723 51553E-01
22	1.34113 44551 43426 29040 88890E-04	1.00812 52768 16658 51393 12795E-01
23	1.24082 86352 52370 43998 17505E-04	1.00817 98491 84650 95408 02947E-01
24	1.15182 49582 53035 64301 67861E-04	1.00822 77583 54343 76933 36687E-01
25	1.07245 56897 48139 28100 22263E-04	1.00827 00460 16238 56120 61623E-01
26	1.00135 29260 81233 48769 06316E-04	1.00830 75580 63334 14974 50412E-01
27	9.37385 73692 79339 02816 25974E-05	1.00834 09871 31499 54546 92214E-01
28	8.79612 22961 73059 68251 62569E-05	1.00837 09047 38696 32309 69970E-01
29	8.27242 60715 59833 96602 10531E-05	1.00839 77858 30067 45607 29498E-01
30	7.79610 42079 23276 25482 67851E-05	1.00842 20277 11833 28116 88292E-01
31	7.36149 99688 63152 54181 05640E-05	1.00844 39647 93074 94535 53285E-01
32	6.96378 56273 93910 50760 41378F-05	1.00846 38801 63256 41160 34668E-01
33	6.59881 98208 97016 51843 27299E-05	1.00848 20147 58346 93354 15393E-01
34	6.26303 28775 28148 72174 88560E-05	1.00849 85746 72774 93649 83798E-01
35	5.95333 39952 36669 72314 26266E-05	1.00851 37370 33693 88865 41556E-01
36	5.66703 56240 88379 11669 53306E-05	1.00852 76547 51702 73374 78452E-01
37	5.40179 14894 07367 48303 17181E-05	1.00854 04603 87015 83099 28634E-01
38	5.15554 55043 87123 80263 84454E-05	1.00855 22693 14381 16465 85801E-01
39	4.92648 94314 06558 12647 19565E-05	1.00856 31823 28403 60397 56105E-01
40	4.71302 76146 35370 71712 84604E-05	1.00857 32877 99538 12006 09865E-01

Table 2.9 The numbers $\left\{ E_n(x^{\frac{5}{4}};[0,1]) \right\}_{n=1}^{40}$

and the products $\left\{ (2n)^{\frac{5}{2}} E_n(x^{\frac{5}{4}};[0,1]) \right\}_{n=1}^{40}$

n	$E_n(x^{\frac{5}{4}};[0,1])$	$(2n)^{\frac{5}{2}} E_n(x^{\frac{5}{4}};[0,1])$
1	4.09600 00000 00000 00000 00000E-02	2.31704 75005 92078 92795 66868E-01
2	8.02524 26337 78619 78009 12148E-03	2.56807 76428 09158 32962 91887E-01
3	3.01196 44802 11402 88926 97435E-03	2.65599 93959 65835 27357 22491E-01
4	1.48803 03515 13865 86732 40839E-03	2.69362 26615 47152 76446 08920E-01
5	8.57820 81604 06160 28000 38316E-04	2.71266 76029 92649 01396 32108E-01
6	5.45979 83042 07701 31043 37602E-04	2.72351 46418 46247 11986 22890E-01
7	3.72289 52467 93096 10164 22305E-04	2.73024 05060 68400 77544 79172E-01
8	2.67059 02530 17198 29007 24304E-04	2.73468 44190 89611 04903 41687E-01
9	1.99166 08993 29419 32812 19724E-04	2.73776 81075 26424 25033 10375E-01
10	1.53170 24993 36673 35655 50320E-04	2.73999 27278 58502 52914 75669E-01
11	1.20768 91129 31596 46410 45988E-04	2.74164 89993 58584 22384 86782E-01
12	9.72040 12600 31670 52294 86484E-05	2.74291 46705 87626 39645 04729E-01
13	7.96040 97604 78546 60333 24509E-05	2.74390 32460 49521 30346 82415E-01
14	6.61604 12734 65470 43829 41372E-05	2.74468 99001 37942 03187 59938E-01
15	5.56917 39757 42218 44017 28778E-05	2.74532 59918 66267 64645 58935E-01
16	4.74025 25154 09923 32022 68018E-05	2.74584 75607 51469 48297 27765E-01
17	4.07424 92968 05693 09157 30007E-05	2.74628 04915 81717 02358 27702E-01
18	3.53220 64803 67711 98660 29427E-05	2.74664 37591 33932 84078 24483E-01
19	3.08597 21632 14778 09140 91901E-05	2.74695 15262 66161 86903 04465E-01
20	2.71482 97354 96205 06433 02821E-05	2.74721 45355 90553 44226 05113E-01
21	2.40328 43447 63621 14994 14606E-05	2.74744 10537 45450 19837 48854E-01
22	2.13957 61483 68854 11172 18301E-05	2.74763 75252 43280 07291 76331E-01
23	1.91466 13376 68234 74588 61520E-05	2.74780 90335 76377 19916 53269E-01
24	1.72150 09040 34394 53416 61088E-05	2.74795 96318 27960 67366 48629E-01
25	1.55455 59207 93585 77724 26252E-05	2.74809 25833 17104 90740 03358E-01
26	1.40942 39324 65847 25679 01904E-05	2.74821 05391 86619 19146 26873E-01
27	1.28257 33589 28201 02554 93888E-05	2.74831 56711 21484 08513 27459E-01
28	1.17114 70166 88517 84638 22362E-05	2.74840 97716 93465 63968 29126E-01
29	1.07281 50604 78442 05135 93551E-05	2.74849 43310 46976 68468 65310E-01
30	9.85663 71708 61017 23072 22245E-06	2.74857 05960 89467 06207 80429E-01
31	9.08110 25074 43540 64186 25047E-06	2.74863 96166 01376 19825 79554E-01
32	8.38837 36617 01329 99424 78624E-06	2.74870 22814 66291 81251 51395E-01
33	7.76742 16347 76041 27731 51666E-06	2.74875 93473 67509 76654 54974E-01
34	7.20896 09209 31234 41355 92033E-06	2.74881 14616 88764 54549 69787E-01
35	6.70513 29307 46223 27461 19951E-06	2.74885 91809 18793 06521 07918E-01
36	6.24925 39284 32718 63341 30767E-06	2.74890 29855 40612 08609 18630E-01
37	5.83561 29794 25635 31780 96580E-06	2.74894 32921 52144 78381 21106E-01
38	5.45930 89543 34390 67178 01901E-06	2.74898 04633 91111 55794 12428E-01
39	5.11611 82491 80383 99976 33394E-06	2.74901 48161 07141 64205 22954E-01
40	4.80238 68232 86260 80710 63098E-06	2.74904 66281 26041 38170 63039E-01

Table 2.10 The numbers $\left\{ E_n(x^{\frac{3}{2}};[0,1]) \right\}_{n=1}^{40}$

and the products $\left\{ (2n)^3 E_n(x^{\frac{3}{2}};[0,1]) \right\}_{n=1}^{40}$

n	$E_n(x^{\frac{3}{2}};[0,1])$	$(2n)^3 E_n(x^{\frac{3}{2}};[0,1])$
1	7.40740 74074 07407 40740 74074E-02	5.92592 59259 25925 92592 59259E-01
2	8.88347 64831 84405 50105 54526E-03	5.68542 49492 38019 52067 54897E-01
3	2.67294 65836 77282 44641 24325E-03	5.77356 46207 42930 08425 08543E-01
4	1.13749 83951 66505 22830 21751E-03	5.82399 17832 52506 76890 71366E-01
5	5.85203 18044 88704 15243 17104E-04	5.85203 18044 88704 15243 17104E-01
6	3.39625 27666.49120 08524 98039E-04	5.86872 47807 69679 50731 16611E-01
7	2.14261 46606 41863 00454 88048E-04	5.87933 46288 01272 08448 19204E-01
8	1.43712 22662 89229 04230 92578E-04	5.88645 28027 20682 15729 87199E-01
9	1.01019 24793 71104 77055 55952E-04	5.89144 25396 92283 02188 02313E-01
10	7.36883 47218 05607 28201 38835E-05	5.89506 77774 44485 82561 11068E-01
11	5.53886 24493 17342 67299 85824E-05	5.89778 07360 33106 47820 88905E-01
12	4.26783 99074 57107 38942 57451E-05	5.89986 18880 68705 25514 21500E-01
13	3.35769 92629 86524 12672 84812E-05	5.90149 22246.25114 80513 79785E-01
14	2.68895 43429 69042 64373 89799E-05	5.90279 25736 85642 41153 58087E-01
15	2.18660 96281 53936 80312 99136E-05	5.90384 59960 15629 36845 07668E-01
16	1.80197 48116 40924 78261 23081E-05	5.90471 10627 84982 32766 40112E-01
17	1.50250 10197 87793 61165 22997E-05	5.90543 00081 73944 01123 81988E-01
18	1.26586 80324 20383 83292 76156E-05	5.90603 38920 60542 81090 70832E-01
19	1.07642 25761 58226 04541 59826E-05	5.90654 59598 95417 95640 65797E-01
20	9.22966 23174 53926 95488 99538E-06	5.90698 38831 70513 25112 95704E-01
21	7.97343 87300 83821 47754 21978E-06	5.90736 12863 44501 65628 14635E-01
22	6.93520 94384 90618 55193 86693E-06	5.90768 88080 83848 50728 34361E-01
23	6.06967 08844 04173 61500 86294E-06	5.90797 48520 43646 42990 47995E-01
24	5.34236 30340 70336 50809 30531E-06	5.90822 61266 39066 55103 02693E-01
25	4.72675 84310 86558 50254 98482E-06	5.90844 80388 58198 12818 73103E-01
26	4.20221 11014 15198 02059 53643E-06	5.90864 49854 77881 63279 87298E-01
27	3.75248 98205 88424 49771 10532E-06	5.90882 05710 91356 75107 57328E-01
28	3.36471 49309 93427 46486 26421E-06	5.90897 77732 13417 57669 31775E-01
29	3.02857 79801 52254 43029 94607E-06	5.90911 90686 34666 66404 58837E-01
30	2.73576 22829 02938 10013 79673E-06	5.90924 65310 70346 29629 80093E-01
31	2.47950 80340 18100 42194 23109E-06	5.90936 19073 14658 37360 66708E-01
32	2.25428 26374 57929 44504 45466E-06	5.90946 66771 37714 56441 75762E-01
33	2.05552 84597 94394 17254 26248E-06	5.90956 21007 70491 47029 31447E-01
34	1.87946 81383 67024 78071 28141E-06	5.90964 92568 30203 35849 09158E-01
35	1.72295 30824 55467 27544 67541E-06	5.90972 90728 22252 75478 23666E-01
36	1.58334 46796 32020 08638 49727E-06	5.90980 23498 32922 33203 01828E-01
37	1.45842 04742 66788 63667 18504E-06	5.90986 97826 42851 58506 71391E-01
38	1.34629 95644 87830 45088 07279E-06	5.90993 19762 06098 60005 81842E-01
39	1.24538 28999 19491 48175 13077E-06	5.90998 94592 25945 21648 04658E-01
40	1.15430 52139 44783 08506 20711E-06	5.91004 26953 97289 39551 78039E-01

Table 2.11 The numbers $\left\{E_n(x^{\frac{7}{4}};[0,1])\right\}_{n=1}^{40}$
and the products $\left\{(2n)^{\frac{7}{2}}E_n(x^{\frac{7}{4}};[0,1])\right\}_{n=1}^{40}$

n	$E_n(x^{\frac{7}{4}};[0,1])$	$(2n)^{\frac{7}{2}}E_n(x^{\frac{7}{4}};[0,1])$
1	1.01611 41224 89277 68285 83569E-01	1.14960 18983 54538 27744 03564E 00
2	5.63623 07912 92347 11510 36143E-03	7.21437 54128 54204 30733 26262E-01
3	1.32786 42459 61522 50259 28594E-03	7.02559 40766 29139 94189 85374E-01
4	4.83302 38522 98617 57411 85654E-04	6.99896 61482 94007 92801 59680E-01
5	2.21205 99166 19973 24396 82554E-04	6.99514 76572 81269 27030 05429E-01
6	1.16867 30496 71406 71758 66473E-04	6.99563 89997 59136 20371 08707E-01
7	6.81484 99779 58745 61771 75425E-05	6.99687 99836 85161 00572 95665E-01
8	4.27129 25430 00665 83781 33779E-05	6.99808 57024 52290 90867 34383E-01
9	2.82871 03581 42012 56563 06296E-05	6.99910 08066 70952 08681 20916E-01
10	1.95653 80363 84233 30810 05859E-05	6.99992 32798 70564 03065 27544E-01
11	1.40169 94989 77925 20672 23154E-05	7.00058 44821 93830 30618 93545E-01
12	1.03377 97896 28476 23574 64135E-05	7.00111 77734 93310 86786 41727E-01
13	7.81245 53534 61452 20273 06987E-06	7.00155 11572 10406 32646 22677E-01
14	6.02786 06113 83805 09140 09854E-06	7.00190 65595 01791 07063 68211E-01
15	4.73489 29736 57098 45446 92094E-06	7.00220 07604 20709 13711 31746E-01
16	3.77767 74847 95661 34402 85460E-06	7.00244 65393 51006 69071 35285E-01
17	3.05552 91945 92718 04757 07084E-06	7.00265 36583 07337 46290 93965E-01
18	2.50158 23703 18365 71196 14876E-06	7.00282 96241 74420 23943 65098E-01
19	2.07033 35208 33333 09148 55306E-06	7.00298 02576 80883 34493 04615E-01
20	1.73013 89578 44698 22301 30284E-06	7.00311 01124 85491 94504 05757E-01
21	1.45856 37835 74176 23130 58190E-06	7.00322 27831 72980 26242 41427E-01
22	1.23942 33489 41085 51197 89707E-06	7.00332 11323 47459 46097 56726E-01
23	1.06085 73349 21207 75680 90783E-06	7.00340 74590 95964 80336 89164E-01
24	9.14049 65008 80109 58541 70479E-07	7.00348 36249 00313 99029 75295E-01
25	7.92361 36152 04385 43589 28904E-07	7.00355 11485 16345 18360 33828E-01
26	6.90733 19978 78772 93732 67229E-07	7.00361 12780 70535 96492 50248E-01
27	6.05267 29541 03888 53783 96519E-07	7.00366 50463 03814 08375 29086E-01
28	5.32930 11972 07404 95249 35537E-07	7.00371 33132 49160 75625 73391E-01
29	4.71338 67827 42933 80051 62789E-07	7.00375 67994 63194 59736 49867E-01
30	4.18604 71762 95607 56879 04577E-07	7.00379 61120 99995 45186 78004E-01
31	3.73219 90018 32778 07605 39392E-07	7.00383 17655 18377 01024 29228E-01
32	3.33970 27004 63453 24547 68298E-07	7.00386 41976 82331 90065 82245E-01
33	2.99871 90376 46803 51914 67544E-07	7.00389 37833 00234 53254 84080E-01
34	2.70122 04859 96829 17725 52041E-07	7.00392 08444 17955 97309 32810E-01
35	2.44061 69766 39379 64384 28420E-07	7.00394 56590 10741 84823 21068E-01
36	2.21146 69062 45458 43586 94868E-07	7.00396 84679 91903 25250 61039E-01
37	2.00925 22628 18333 57867 47594E-07	7.00398 94809 61273 28051 68507E-01
38	1.83020 23771 88989 46024 92260E-07	7.00400 88809 44557 25396 93949E-01
39	1.67115 48416 69500 30553 07923E-07	7.00402 68283 20086 34855 30513E-01
40	1.52944 50491 27260 76307 30793E-07	7.00404 34640 87675 83603 53259E-01

Table 2.12 The numbers $\left\{ E_n(x^{\frac{9}{4}};[0,1]) \right\}_{n=1}^{40}$

and the products $\left\{ (2n)^{\frac{9}{2}} E_n(x^{\frac{9}{4}};[0,1]) \right\}_{n=1}^{40}$

n	$E_n(x^{\frac{9}{4}};[0,1])$	$(2n)^{\frac{9}{2}} E_n(x^{\frac{9}{4}};[0,1])$
1	1.45194 94105 24288 36121 60220E$-$01	3.28538 64771 88910 82721 21530E$+$00
2	7.01508 71176 00410 94333 14408E$-$03	3.59172 46042 11410 40298 56977E$+$00
3	8.85497 56243 15935 79310 91840E$-$04	2.81104 62865 80661 16859 57090E$+$00
4	2.28108 63185 39960 25202 60709E$-$04	2.64269 26765 03900 84705 70912E$+$00
5	8.15287 86027 10627 99019 50781E$-$05	2.57816 65871 41661 05909 57083E$+$00
6	3.54502 18457 45929 87008 37030E$-$05	2.54644 63053 44806 56540 33791E$+$00
7	1.75904 07104 34746 05339 19473E$-$05	2.52843 65008 75205 19051 25513E$+$00
8	9.60234 13406 33154 55863 52816E$-$06	2.51719 61683 98937 66861 88873E$+$00
9	5.63500 82388 81366 15972 01515E$-$06	2.50969 43232 20262 85157 17138E$+$00
10	3.50004 82060 64100 65803 76949E$-$06	2.50443 06282 51366 63074 20694E$+$00
11	2.27583 42548 97137 17829 65734E$-$06	2.50059 11721 51351 75300 00247E$+$00
12	1.53670 35771 94084 23573 42077E$-$06	2.49770 23930 60108 63330 23757E$+$00
13	1.07095 89471 66745 88990 32799E$-$06	2.49547 30802 52353 21916 25233E$+$00
14	7.66718 31999 49634 29523 87603E$-$07	2.49371 59407 74778 58581 92665E$+$00
15	5.61766 33520 18559 52380 61667E$-$07	2.49230 59600 80226 25382 85947E$+$00
16	4.19977 67423 34087 76451 95921E$-$07	2.49115 70441 90705 03188 97415E$+$00
17	3.19580 70429 11614 42502 63913E$-$07	2.49020 82993 55687 55775 07455E$+$00
18	2.47022 30066 62623 14504 05300E$-$07	2.48941 56513 35189 06182 82369E$+$00
19	1.93621 90397 25666 05504 22805E$-$07	2.48874 65511 49412 17077 78296E$+$00
20	1.53677 83319 59254 44266 25151E$-$07	2.48817 65313 46214 94438 55849E$+$00
21	1.23359 87304 57646 96904 88767E$-$07	2.48768 69077 13392 37184 44935E$+$00
22	1.00042 53824 09987 21558 19397E$-$07	2.48726 32101 85612 40522 27454E$+$00
23	8.18930 50278 96281 02899 28832E$-$08	2.48689 40896 62178 26767 96279E$+$00
24	6.76106 66346 73287 11515 24094E$-$08	2.48657 05424 15587 65719 61399E$+$00
25	5.62582 15406 67504 66158 66023E$-$08	2.48628 53507 19589 30855 59149E$+$00
26	4.71510 95316 65866 20586 14593E$-$08	2.48603 26733 67397 19459 91304E$+$00
27	3.97828 24787 52731 57920 53005E$-$08	2.48580 77418 19382 33201 59100E$+$00
28	3.37742 93840 32474 29540 63160E$-$08	2.48560 66319 13635 93523 79120E$+$00
29	2.88386 47056 84015 98427 14597E$-$08	2.48542 60903 78928 03846 73617E$+$00
30	2.47566 45486 96487 69533 01735E$-$08	2.48526 34015 99057 26737 25131E$+$00
31	2.13591 57216 97454 83511 99909E$-$08	2.48511 62842 88183 54208 72702E$+$00
32	1.85145 64712 60395 88710 31158E$-$08	2.48498 28106 34676 32347 52470E$+$00
33	1.61196 07014 52640 14239 96232E$-$08	2.48486 13424 93137 28382 74563E$+$00
34	1.40926 50325 21164 11851 96743E$-$08	2.48475 04806 33376 58581 74931E$+$00
35	1.23686 94701 09060 51153 23680E$-$08	2.48464 90240 77203 50758 69492E$+$00
36	1.08956 34971 38373 87343 66257E$-$08	2.48455 59372 93007 70712 93935E$+$00
37	9.63143 67016 33722 03406 58917E$-$09	2.48447 03235 58161 25012 56287E$+$00
38	8.54198 59451 96151 53165 62591E$-$09	2.48439 14031 97724 70300 05123E$+$00
39	7.59943 94772 41321 80491 21705E$-$09	2.48431 84957 04612 71369 90147E$+$00
40	6.78094 98538 61360 90662 69043E$-$09	2.48425 10049 69167 71467 67864E$+$00

Table 2.13 The numbers $\left\{E_n(x^{\frac{5}{2}};[0,1])\right\}_{n=1}^{40}$

and the products $\left\{(2n)^5 E_n(x^{\frac{5}{2}};[0,1])\right\}_{n=1}^{40}$

n	$E_n(x^{\frac{5}{2}};[0,1])$	$(2n)^5 E_n(x^{\frac{5}{2}};[0,1])$
1	1.62865 05699 56943 94291 08537E-01	5.21168 18238 62220 61731 47318E+00
2	1.47939 44809 96337 72829 42593E-02	1.51489 99485 40249 83377 33215E+01
3	1.18214 04692 91244 23479 59960E-03	9.19232 42892 08715 16977 36651E+00
4	2.49828 08358 39482 67178 31553E-04	8.18636 66428 78816 81889 90432E+00
5	7.81682 53492 17701 19637 23277E-05	7.81682 53492 17701 19637 23277E+00
6	3.06930 57765 27008 29312 48889E-05	7.63741 49498 47685 27594 85235E+00
7	1.40120 06484 36899 81379 73831E-05	7.53599 33754 49272 05455 76376E+00
8	7.12661 26018 89121 67231 48650E-06	7.47279 49356 38487 64666 92319E+00
9	3.93245 39829 57197 51397 09442E-06	7.43063 92076 68465 79207 90490E+00
10	2.31283 25568 58472 45740 74790E-06	7.40106 41819 47111 86370 39330E+00
11	1.43190 09232 61686 21769 27725E-06	7.37949 04189 50970 46546 04383E+00
12	9.24727 39618 48993 20552 00483E-07	7.36325 65583 19387 76741 10869E+00
13	6.18676 39825 99188 09881 22754E-07	7.35072 69100 51841 10967 13797E+00
14	4.26536 47458 13107 38645 47909E-07	7.34084 96929 67003 73444 05167E+00
15	3.01766 37387 09866 35656 28211E-07	7.33292 88850 64975 24644 76553E+00
16	2.18345 61213 38618 45028 77496E-07	7.32646 29948 44042 17641 25675E+00
17	1.61132 60031 70660 52589 86643E-07	7.32112 80156 28430 53542 68795E+00
18	1.21004 35085 38399 80756 35146E-07	7.31667 03754 94038 55225 01604E+00
19	9.22937 04955 63383 01297 57585E-08	7.31290 72175 02079 07681 48041E+00
20	7.13837 99491 28881 25499 25701E-08	7.30970 10679 07974 40511 23918E+00
21	5.59100 01167 43914 46899 48648E-08	7.30694 69336 94060 10455 56469E+00
22	4.42925 70590 02669 13082 67557E-08	7.30456 34929 60653 98977 31055E+00
23	3.54553 38131 42084 38003 59566E-08	7.30248 69566 33716 09755 32061E+00
24	2.86520 91920 95143 43040 44150E-08	7.30066 67129 59167 79596 17678E+00
25	2.33569 98994 72922 44495 01155E-08	7.29906 21858 52882 64046 91110E+00
26	1.91940 11427 05985 75400 57487E-08	7.29764 05348 22231 74207 54579E+00
27	1.58905 28604 41589 90251 78967E-08	7.29637 49480 19312 78187 78768E+00
28	1.32464 54404 04562 84923 86095E-08	7.29524 33596 43070 56464 79964E+00
29	1.11132 05242 57453 80960 09435E-08	7.29422 74751 36879 82378 96261E+00
30	9.37925 92881 69741 61248 25676E-09	7.29331 20224 80791 07786 64446E+00
31	7.96007 29466 19499 05295 85864E-09	7.29248 41715 13106 49380 82677E+00
32	6.79095 56249 94476 64542 33090E-09	7.29173 30794 84629 34318 22251E+00
33	5.82197 54657 26465 40522 64928E-09	7.29104 95324 02024 13230 21776E+00
34	5.01427 70070 09186 19891 65702E-09	7.29042 56597 41227 09896 71266E+00
35	4.33739 19830 13860 82815 60303E-09	7.28985 47058 51395 89388 18401E+00
36	3.76725 64066 64874 42383 86980E-09	7.28933 08455 20827 83775 13379E+00
37	3.28473 51402 09095 70695 46648E-09	7.28884 90342 09552 11878 23642E+00
38	2.87451 46685 10051 17214 93374E-09	7.28840 48856 91462 85804 31894E+00
39	2.52426 54729 45432 18910 33143E-09	7.28799 45715 15449 27906 12179E+00
40	2.22400 35211 01882 76758 41368E-09	7.28761 47379 46649 45281 96996E+00

Table 2.14 The numbers $\left\{ E_n(x^{\frac{11}{4}};[0,1]) \right\}_{n=1}^{40}$

and the products $\left\{ (2n)^{\frac{11}{2}} E_n(x^{\frac{11}{4}};[0,1]) \right\}_{n=1}^{40}$

n	$E_n(x^{\frac{11}{4}};[0,1])$	$(2n)^{\frac{11}{2}} E_n(x^{\frac{11}{4}};[0,1])$
1	1.78495 84271 97176 77604 05261E$-$01	8.07779 97312 46067 82813 46075E$+$00
2	2.29531 13865 66331 29850 34158E$-$02	4.70079 77196 87846 49933 49957E$+$01
3	8.73213 75344 52217 73395 38885E$-$04	1.66323 05156 93019 38060 91796E$+$01
4	1.47477 78542 25843 24210 11159E$-$04	1.36685 21364 25015 76416 26342E$+$01
5	4.00493 11874 48502 11678 76009E$-$05	1.26647 04424 58001 82824 24256E$+$01
6	1.41435 91657 69780 60681 91415E$-$05	1.21914 83704 71219 35975 29151E$+$01
7	5.92723 88070 58846 95438 22275E$-$06	1.19276 97639 04708 92312 54170E$+$01
8	2.80489 97221 84789 33398 43727E$-$06	1.17646 02124 35855 06426 87990E$+$01
9	1.45399 55666 10093 33142 00412E$-$06	1.16563 30703 82115 72062 39159E$+$01
10	8.09219 74867 51948 04158 32355E$-$07	1.15806 10347 34739 81973 36942E$+$01
11	4.76798 51587 04478 95848 44949E$-$07	1.15254 96400 50862 37603 78384E$+$01
12	2.94397 93256 35154 04975 74904E$-$07	1.14840 89942 99575 38214 80365E$+$01
13	1.89031 57477 80923 87764 58369E$-$07	1.14521 69472 07390 79925 21057E$+$01
14	1.25477 01191 63224 15027 87156E$-$07	1.14270 29283 60980 67422 59603E$+$01
15	8.57036 87928 83018 26456 56236E$-$08	1.14068 67883 02085 06451 66260E$+$01
16	6.00089 30215 89004 85683 57027E$-$08	1.13904 46941 79287 20695 49428E$+$01
17	4.29427 23596 78115 73805 07690E$-$08	1.13768 91842 65426 61638 56524E$+$01
18	3.13276 25350 59799 88040 07694E$-$08	1.13655 70248 66792 18055 85513E$+$01
19	2.32496 38178 78564 17558 09414E$-$08	1.13560 15587 79560 27694 63077E$+$01
20	1.75220 40592 54358 34264 66382E$-$08	1.13478 77381 39936 47997 11412E$+$01
21	1.33898 57956 55380 64930 02694E$-$08	1.13408 88147 12662 33070 13669E$+$01
22	1.03615 73379 82858 51342 65663E$-$08	1.13348 40831 92020 60736 92633E$+$01
23	8.11045 97226 07716 34104 92046E$-$09	1.13295 73106 32464 30391 53912E$+$01
24	6.41519 30011 98588 88259 59444E$-$09	1.13249 56231 40934 80221 98436E$+$01
25	5.12324 86346 94376 86572 86691E$-$09	1.13208 87035 30348 46688 07845E$+$01
26	4.12784 84983 31339 85943 34433E$-$09	1.13172 82043 38673 29078 96242E$+$01
27	3.35315 24760 73564 99473 03792E$-$09	1.13140 73124 97929 53268 02203E$+$01
28	2.74457 24218 26856 64190 72267E$-$09	1.13112 04224 13227 24814 58376E$+$01
29	2.26233 01696 02348 32582 42787E$-$09	1.13086 28876 20620 03382 87675E$+$01
30	1.87710 66425 76869 19453 16864E$-$09	1.13063 08301 30300 41878 47711E$+$01
31	1.56706 10703 10308 98864 77024E$-$09	1.13042 09926 16945 20745 23101E$+$01
32	1.31576 16168 99836 85708 04919E$-$09	1.13023 06227 83376 04177 92277E$+$01
33	1.11073 08930 87327 12193 36322E$-$09	1.13005 73821 27563 10424 94712E$+$01
34	9.42411 78734 76660 25172 54428E$-$10	1.12989 92733 94072 57680 61518E$+$01
35	8.03424 16759 65144 56829 52801E$-$10	1.12975 45824 57186 73262 95549E$+$01
36	6.88025 20236 46813 24464 27064E$-$10	1.12962 18314 42662 74133 69997E$+$01
37	5.91713 77135 34796 81229 32044E$-$10	1.12949 97406 69101 60493 30468E$+$01
38	5.10937 88136 55715 63028 98748E$-$10	1.12938 71975 60751 13235 28720E$+$01
39	4.42876 42679 31851 40722 64057E$-$10	1.12928 32311 08459 46816 10333E$+$01
40	3.85275 09623 41573 01881 33378E$-$10	1.12918 69907 74483 62845 59615E$+$01

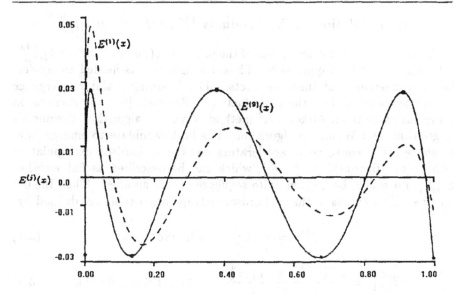

Figure 2.1 The error curves $E^{(1)}(x)$ and $E^{(9)}(x)$ for
$x^{\frac{1}{2}}$ on $[0,1]$ and $n = 5$

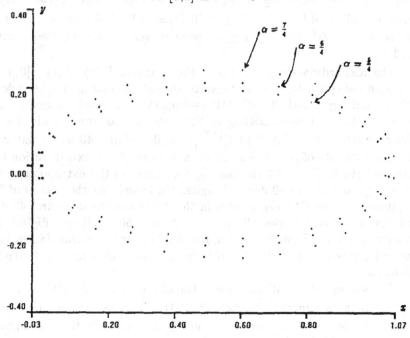

Figure 2.2 Zeros of the best uniform polynomial approximants
to x^{α} on $[0,1]$ for $n = 30$ and $\alpha = \frac{5}{4}$, $\frac{6}{4}$, and $\frac{7}{4}$

3. Extrapolation of the Products $\left\{(2n)^{2\alpha}E_n(x^\alpha;[0,1])\right\}_{n=1}^{40}$

As can be seen, the convergence of the products $\left\{(2n)^{2\alpha}E_n(x^\alpha;[0,1])\right\}_{n=1}^{40}$ in Tables 2.2 – 2.14 is quite slow. Thus, an algorithm is needed to accelerate the convergence of these products. The following scalar convergence accelerators were used on these products (cf. Brezinski [6]): the Richardson extrapolation method, Aitken's Δ^2 method, Wynn's ε algorithm, Brezinski's θ algorithm, and Wynn's ρ algorithm. The best candidate to emerge from the above list of convergence accelerators was the Richardson extrapolation method (cf. Brezinski [6, p. 6]), which can be described as follows: Let $\{S_i\}_{i=1}^n$, for $n \geq 2$, be a given finite sequence of real numbers. Then the 0th and $(k+1)^{th}$ columns of the Richardson extrapolation table are defined by

$$T_0^{(i)} := S_i \ (1 \leq i \leq n), \text{ and} \tag{3.1}$$

$$T_{k+1}^{(i)} := \frac{x_i T_k^{(i+1)} - x_{i+k+1} T_k^{(i)}}{x_i - x_{i+k+1}} \qquad (1 \leq i \leq n - k - 1), \tag{3.2}$$

for each $k = 0, 1, \cdots, n-2$, where $\{x_i\}_{i=1}^n$ are given constants. In this way, a triangular table, consisting of $\frac{1}{2}n(n+1)$ entries, is created. Preliminary calculations indicated that $(2i)^{2\alpha}E_i(x^\alpha;[0,1]) \approx \beta(\alpha) + \frac{k(\alpha)}{i^2} +$ lower-order terms, and so $x_i := 1/i^2$, for $1 \leq i \leq n$, were chosen as the required constants in (3.2).

The Richardson extrapolation of the products $\left\{(2i)^{2\alpha}E_i(x^\alpha;[0,1])\right\}_{i=1}^{40}$ produced unexpectedly beautiful results, the calculations having been done to 200 decimal digits with Brent's MP package [5]. The resulting extrapolation table had 40 columns consisting of 820 entries. Column 1 had 40 entries (the products $\left\{(2i)^{2\alpha}E_i(x^\alpha;[0,1])\right\}_{i=1}^{40}$), while column 40 had 1 entry (i.e., the final estimate of $\beta(\alpha)$). Rather than present all the extrapolation tables here, we give in Table 3.1 the last eight columns of the extrapolation table for $\alpha = \frac{3}{8}$, rounded to 80 decimal digits. On examining the entries of Table 3.1, it can be seen that *every* entry in this table has the same first 42 digits, and that the latter entries all agree to at least 50 significant digits! More importantly, what Table 3.1 shows, for the Richardson extrapolation for the particular case $\alpha = \frac{3}{8}$, actually holds for *all* values of α for which $\beta(\alpha)$ was calculated.

The success of the Richardson extrapolation method (with $x_i := 1/i^2$) applied to the products $\left\{(2i)^{2\alpha}E_i(x^\alpha;[0,1])\right\}_{i=1}^{40}$ gives strong numerical evidence for the following new conjecture (which extends the corresponding conjecture in [18] for the case $\alpha = \frac{1}{2}$):

CONJECTURE: $(2i)^{2\alpha}E_i(x^\alpha;[0,1])$ *admits an asymptotic series expansion of the form*

$$(2i)^{2\alpha}E_i(x^\alpha;[0,1]) \approx \beta(\alpha) - \frac{K_1(\alpha)}{i^2} + \frac{K_2(\alpha)}{i^4} - \frac{K_3(\alpha)}{i^6} + \cdots, \qquad (3.3)$$

as $i \to \infty$, *where the constants* $K_j(\alpha)$ $(j = 1, 2, \cdots)$ *are independent of* i.
If we assume that (3.3) is valid, it follows that

$$i^2\left((2i)^{2\alpha}E_i(x^\alpha;[0,1]) - \beta(\alpha)\right) \approx -K_1(\alpha) + \frac{K_2(\alpha)}{i^2} - \frac{K_3(\alpha)}{i^4} + \cdots, \text{ as } i \to \infty.$$
$$(3.4)$$

We can apply Richardson extrapolation to the sequence

$$\left\{i^2\left((2i)^{2\alpha}E_i(x^\alpha;[0,1]) - \beta(\alpha)\right)\right\}_{i=1}^{40},$$

with $x_i = 1/i^2$, to obtain an extrapolated estimate for $K_1(\alpha)$ of (3.4). This boot-strapping procedure can be continued to give, via Richardson extrapolations, extrapolated estimates for the successive $K_j(\alpha)$'s of (3.3). In Table 3.2, extrapolated estimates of $\{K_j(\alpha)\}_{j=1}^{40}$ are given, rounded to 25 decimal digits, for the two cases $\alpha = \frac{1}{4}$ and $\alpha = \frac{1}{2}$. These estimates were computed to 200 decimal digits with Brent's MP package [5].

It turns out, for the cases we considered, that the constants $\{K_j(\alpha)\}_{j=1}^{40}$ in (3.3) are all *positive* for $\alpha \in (0,2)$ with $\alpha \neq 1$. There are differences when $\alpha > 2$. For example, when $\alpha = \frac{9}{4}$, the constant $K_1(\frac{9}{4})$ is negative, while the constants $\{K_j(\frac{9}{4})\}_{j=2}^{40}$ are all positive.

With the values of $\{K_j(\frac{1}{2})\}_{j=1}^{40}$ from the right column of Table 3.2, we wish to show now how these numbers can be used in (3.3) for $\alpha = \frac{1}{2}$, to *estimate numerically* the numbers $\left\{2nE_n(x^{\frac{1}{2}};[0,1])\right\}_{n=1}^{10}$ of Table 2.5. Because the numbers $\{K_j(\frac{1}{2})\}_{j=1}^{40}$ are all positive in Table 3.2, then what is conjectured to be an *asymptotic series* in (3.3) for $\alpha = \frac{1}{2}$, namely

$$2nE_n(x^{\frac{1}{2}};[0,1]) \approx \beta(\tfrac{1}{2}) - \frac{K_1(\frac{1}{2})}{n^2} + \frac{K_2(\frac{1}{2})}{n^4} - \frac{K_3(\frac{1}{2})}{n^6} + \cdots, \text{ as } n \to \infty, \quad (3.5)$$

is an *alternating* asymptotic series. The general rule used in summing such an alternating asymptotic series is (cf. Gradshteyn and Ryzhik [9, p. 18]) to add the terms, up to and including the *smallest* term in modulus, and discarding all subsequent terms in (3.5). In this case, the error (for a true alternating asymptotic series) is bounded above by the modulus of the first discarded term. This method of adding the alternating series in (3.5) was applied for each n with $n = 1, 2, \cdots, 10$, and in Table 3.3, we give, for each

n, the numerical estimates of $\left\{2nE_n(x^{\frac{1}{2}};[0,1])\right\}_{n=1}^{10}$, the number of terms added in (3.5), and the resulting number of digits of $\left\{2nE_n(x^{\frac{1}{2}};[0,1])\right\}_{n=1}^{10}$ from Table 2.5, which are correctly given by this sum.

The accuracy of the conjectured asymptotic series in (3.5), as judged by the last column of Table 3.3, is surprisingly good, even for very low values of n. We believe that this tends to give further *credence* to our conjecture in (3.3).

Acknowledgments. A very special thanks goes to Dr. Ewing L. Lusk of the Mathematics and Computer Science Division, Argonne National Laboratory, for introducing the first author to parallel processing and for teaching him how to use the p4 package. The Advanced Computing Research Facility at Argonne provided the ideal environment in which these computations were performed.

Table 3.1 The last eight columns of the extrapolation table for $\alpha = \frac{3}{8}$

Column 33
0.3152412741461107187646673856654824993299382682493996953174161125239567 1501432981
0.3152412741461107187646673856654824993299383300136464677619789380900618 3093591420
0.3152412741461107187646673856654824993299383329813874770364250736440757 4134051357
0.3152412741461107187646673856654824993299383331304255993064100975045924 0994018703
0.3152412741461107187646673856654824993299383331383992663548111118126824 3245197816
0.3152412741461107187646673856654824993299383331388590227546789069684871 3932699485
0.3152412741461107187646673856654824993299383331388877669470226980349865 5613444183
0.3152412741461107187646673856654824993299383331388897201876454171473889 3161164126

Column 34
0.3152412741461107187646673856654824993299383300671220060930997717052705 7995402813
0.3152412741461107187646673856654824993299383329911098063051374278965049 4334020104
0.3152412741461107187646673856654824993299383331314678239376687340350855 3349682811
0.3152412741461107187646673856654824993299383331384935595349843684489939 1830577613
0.3152412741461107187646673856654824993299383331388671227617237941276803 1598088873
0.3152412741461107187646673856654824993299383331388884637759280020608410 8745098600
0.3152412741461107187646673856654824993299383331388897818954407190406885 5533722654

Column 35
0.3152412741461107187646673356654824993299383329934986852268792887267265 4004231393
0.3152412741461107187646673856654824993299383331319023688838994811129139 5637656999
0.3152412741461107187646673856654824993299383331385400533734960160296741 9438171770
0.3152412741461107187646673856654824993299383331388713083440962246674975 3052066590
0.3152412741461107187646673856654824993299383331388888204105236606158738 5435389538
0.3152412741461107187646673856654824993299383331388898122357875590734727 6815302236

Column 36
0.3152412741461107187646673856654824993299383331320092143152948635008569 5824246516
0.3152412741461107187646673856654824993299383331355595044635754197949980 3390700751
0.3152412741461107187646673856654824993299383331388733859014031249307312 6572989972
0.3152412741461107187646673856654824993299383331388890065853162446525343 3354414898
0.3152412741461107187646673856654824993299383331388898279790457161918473 5408634183

Column 37
0.3152412741461107187646673856654824993299383331385642926654381979794235 7562898452
0.3152412741461107187646673856654824993299383331388742577942859796672194 1359607442
0.3152412741461107187646673856654824993299383331388890995655776322699260 1847161475
0.3152412741461107187646673856654824993299383331388898362759520744902242 5328373772

Column 38
0.3152412741461107187646673856654824993299383331388744726003350772902661 1791898994
0.3152412741461107187646673856654824993299383331388891387001097130941915 5994788512
0.3152412741461107187646673856654824993299383331388898404433897677523126 7824496280

Column 39
0.3152412741461107187646673856654824993299383331388891483488595648283133 7898869361
0.3152412741461107187646673856634824993299383331388898422021448556085986 9608530636

Column 40
0.3152412741461107187646673856654824993299383331388898426360743648586176 3624609223

Table 3.2 The numbers $\{K_j(\alpha)\}_{j=1}^{40}$ for the products

$$\left\{(2n)^{\frac{1}{2}} E_n(x^{\frac{1}{4}};[0,1])\right\}_{n=1}^{40} \text{ and } \left\{2nE_n(x^{\frac{1}{2}};[0,1])\right\}_{n=1}^{40}$$

j	$K_j(\frac{1}{4})$ for $(2n)^{\frac{1}{2}} E_n(x^{\frac{1}{4}};[0,1])$	$K_j(\frac{1}{2})$ for $2nE_n(x^{\frac{1}{2}};[0,1])$
1	1.88287 39793 84817 89228 09318E-02	4.39675 28880 37559 56907 22422E-02
2	7.00185 85948 43410 47450 01816E-03	2.64071 68775 63975 40815 39953E-02
3	5.73516 97516 32046 08075 86193E-03	3.12534 26468 88366 21604 92812E-02
4	8.06896 10339 10935 34854 06498E-03	5.88900 16571 94464 07344 76045E-02
5	1.74579 86898 22971 29737 14933E-02	1.60106 99716 56970 12802 60793E-01
6	5.43809 24778 10764 14890 08586E-02	5.95435 31510 38536 37996 70917E-01
7	2.32493 06747 75236 92465 30680E-01	2.92591 54709 09529 94602 95871E+00
8	1.31310 12172 46725 96283 60417E+00	1.84941 40338 08461 76530 65579E+01
9	9.49486 35907 59470 93940 69982E+00	1.46943 01234 81390 08073 36135E+02
10	8.56541 78912 67836 39867 72242E+01	1.43803 27177 19218 36274 50095E+03
11	9.43599 97850 93765 79285 94514E+02	1.70262 52451 48179 70302 60874E+04
12	1.24700 85029 18428 51920 24258E+04	2.40118 63088 02885 85761 84991E+05
13	1.94752 59148 98658 72577 90066E+05	3.97938 55356 34528 46153 52389E+06
14	3.54888 23773 90288 38001 09713E+06	7.65939 47118 46820 15817 55294E+07
15	7.46332 98488 63627 69015 04457E+07	1.69481 50880 32764 21388 41074E+09
16	1.79420 05295 60613 89130 60764E+09	4.27276 88119 78247 93977 76088E+10
17	4.88971 13775 38487 47692 20864E+10	1.21765 42900 93742 38977 76490E+12
18	1.49957 83078 82119 28648 24578E+12	3.89510 79873 27815 60235 55055E+13
19	5.14140 87071 60691 14851 54086E+13	1.38987 13708 56504 27722 89745E+15
20	1.95917 79147 22797 36052 51848E+15	5.50104 16293 72886 50088 08098E+16
21	8.25372 63078 93983 97215 36782E+16	2.40283 18915 18267 43096 78599E+18
22	3.82590 58790 26863 79946 66492E+18	1.15294 25357 75728 36192 51101E+20
23	1.94284 65029 20831 32751 20729E+20	6.05164 23525 26015 18598 53359E+21
24	1.07656 07728 79016 03680 36217E+22	3.46139 35603 27124 64102 04503E+23
25	6.48565 36937 36696 87497 40010E+23	2.14984 92585 89692 81716 58944E+25
26	4.23378 83030 85676 73979 22890E+25	1.44521 08550 86315 35148 11714E+27
27	2.98550 41518 97459 60246 62921E+27	1.04835 78587 43688 17461 00059E+29
28	2.26752 07339 22410 12868 00710E+29	8.18281 06018 33942 82217 61899E+30
29	1.84953 46875 82409 12337 49776E+31	6.85254 13926 52250 61076 32825E+32
30	1.61458 33856 41541 53327 34624E+33	6.13524 03190 25207 75193 57221E+34
31	1.50073 51996 39063 86688 88165E+35	5.84094 79665 76950 20135 61201E+36
32	1.47195 35912 45356 18949 33492E+37	5.85710 21598 72912 73130 62501E+38
33	1.50062 87502 36148 80295 52069E+39	6.08952 24914 19377 29195 01726E+40
34	1.55533 93658 09662 18046 90055E+41	6.41699 96149 92711 87493 01466E+42
35	1.59427 33971 86612 52530 68539E+43	6.66604 10147 94249 78486 15605E+44
36	1.56879 83657 37115 21278 64428E+45	6.62788 48678 47571 75585 59365E+46
37	1.43989 06906 27776 24195 36193E+47	6.13133 94490 51024 25546 54637E+48
38	1.20074 39921 82110 70341 69152E+49	5.14335 38961 72948 37983 87026E+50
39	8.88579 04196 78705 74474 36614E+50	3.82318 47580 49471 00098 49618E+52
40	5.71045 56570 95263 56043 17223E+52	2.46526 33006 55214 86369 90342E+54

Table 3.3 The numerical estimates of $\left\{2nE_n(x^{\frac{1}{2}};[0,1])\right\}_{n=1}^{10}$

n	Numerical estimate of $2nE_n(x^{\frac{1}{2}};[0,1])$	Number of terms used	Number of correct digits
1	2.62609 13891 91332 91427 25402E-01	2	1
2	2.70558 78465 14674 71053 48747E-01	6	3
3	2.75574 18419 18367 11887 55676E-01	9	6
4	2.77517 82509 28654 28588 84202E-01	12	8
5	2.78451 18553 46228 66724 98914E-01	15	11
6	2.78967 91746 49567 94439 09082E-01	19	14
7	2.79282 94495 85180 28421 41933E-01	22	17
8	2.79488 83759 45074 47709 85560E-01	25	19
9	2.79630 65741 01282 01252 92946E-01	28	22
10	2.79732 43377 19738 29689 89387E-01	31	25

References

[1] S. N. Bernstein, *Sur la meilleure approximation de $|x|$ par des polynômes de degrés donnés*, Acta Math. **37** (1913), pp. 1–57.

[2] S. N. Bernstein, *Sur la meilleure approximation de $|x|^p$ par des polynômes de degrés trés élevés*, Bull. Acad. Sci. USSR, Cl. sci. math. nat. **2** (1938), pp. 181–190.

[3] S. N. Bernstein, *Collected Works* (Russian), Akad. Nauk SSSR, Moscow, Vol. II, 1954, pp. 262–272.

[4] H.-P. Blatt and E. B. Saff, *Behavior of zeros of polynomials of near best approximation*, J. Approx. Theory **46** (1986), pp. 323–344.

[5] R. P. Brent, *A FORTRAN multiple-precision arithmetic package*, ACM Trans. Math. Soft. **4** (1978), pp. 57–70.

[6] C. Brezinski, *Algorithms d'Accélération de la Convergence*, Éditions Technip, Paris, 1978.

[7] A. J. Carpenter and R. S. Varga, *Some numerical results on best uniform rational approximation of x^α on $[0,1]$*, to appear.

[8] A. A. Gonchar and E. A. Rakhmanov, *Equilibrium distributions and degree of rational approximation of analytic functions*, Mat. Sbornik **134**, (176) (1987), pp. 306–352. An English translation appears in Math. USSR Sbornik **62**, 2 (1989), pp. 305–348.

[9] I. S. Gradshteyn and I. M. Ryzhik, *Table of Integrals, Series, and Products, Corrected and Enlarged Edition Prepared by Alan Jeffrey*, Academic Press, San Diego, 1979.

[10] M. A. Jenkins, *Algorithm 493, Zeros of a real polynomial*, Collected Algorithms from ACM, 1975, 10 pp.

[11] H. L. Loeb, *Approximation by generalized rationals*, SIAM J. on Numer. Anal. **3** (1966), pp. 34–55.

[12] E. L. Lusk and R. A. Overbeek, *Use of monitors in FORTRAN: A tutotial on the barrier, self-scheduling do-loop, and askfor monitors*, Parallel MIMD Computation: The HEP Supercomputer and Its Applications (J. S. Kowalik, ed.), pp. 367–411, The MIT Press, Cambridge, Mass., 1985.

[13] G. Meinardus, *Approximation of Functions: Theory and Numerical Methods*, Springer-Verlag, New York, 1967.

[14] E. Ya. Remez, *Sur le calcul effectiv des polynômes d'approximation de Tchebichef*, C.R. Acad. Sci. Paris **199** (1934), pp. 337–340.

[15] T. J. Rivlin, *An Introduction to the Approximation of Functions*, Blaisdell Publishing Co., Waltham, Mass., 1969.

[16] H. Stahl, *Best uniform rational approximation of $|x|$ on $[-1,1]$*, Mat. Sbornik (to appear).

[17] R. S. Varga, *Scientific Computation on Mathematical Problems and Conjectures*, CBMS-NSF Regional Conference Series in Applied Mathematics 60, SIAM, Philadelphia, Penn., 1990.

[18] R. S. Varga and A. J. Carpenter, *On the Bernstein Conjecture in approximation theory*, Constr. Approx. 1 (1985), pp. 333–348. A Russian translation appears in Mat. Sbornik 129, 171 (1986), pp. 535–548.

[19] R. S. Varga, A. Ruttan, and A. J. Carpenter, *Numerical results on best uniform rational approximation of $|x|$ on $[-1, +1]$*, Mat. Sbornik (to appear).

Amos J. Carpenter
Dept. of Math. and Computer Science
Butler University
Indianapolis, IN 46208

Richard S. Varga
Institute for Computational Mathematics
Kent State University
Kent, OH 44242